SOUTH-WESTERN'S

2003 Edition

Payroll Accounting

Bernard J. Bieg

Bucks County Community College

SOUTH-WESTERN

TM

THOMSON LEARNING

Australia · Canada · Mexico · Singapore · Spain · United Kingdom · United States

SOUTH-WESTERN
THOMSON LEARNING

Payroll Accounting, 2003 edition

Bernard J. Bieg

Editor-in-Chief:
Jack W. Calhoun

Team Leader:
Melissa S. Acuña

Acquisitions Editor:
Jennifer L. Codner

Developmental Editor:
Carol Bennett

Marketing Manager:
Mignon Tucker

Senior Production Editor:
Marci Combs

Production Editor–Software:
Robin Browning

Editorial Assistant:
Janice Hughes

Production House:
Litten Editing and Production, Inc.

Compositor:
Parkwood Composition

Manufacturing Coordinator:
Doug Wilke

Printer:
QuebecorWorld

Internal Design:
Ramsdell Design

Cover Design:
Ramsdell Design

Photo Researcher:
Deanna Ettinger

CONTEMPORARY'S
READING
SKILLS THAT WORK
A Functional Approach for Life and Work

BOOK ONE

SUSAN ECHAORE-YOON

Project Editor
Mark Boone

Consultant
Elizabeth Kappel
Program Coordinator
for Workplace Literacy
Triton College
River Grove, IL

CONTEMPORARY BOOKS

a division of NTC/CONTEMPORARY PUBLISHING GROUP
Lincolnwood, Illinois USA

Photo credits: p. 1, 5, 26, 49, 63—© Ralph J. Brunke; p. 2—
© Will McIntyre/Photo Researchers; p. 6—© Bohdan
Hrynewych/Stock, Boston; p. 10—© Paul Fusco/Magnum;
p. 16, 33—© Robert Frerck/Odyssey Productions; p. 18—©
Charles Gupton/Tony Stone Worldwide from Click,
Chicago; p. 20—© Phyllis Graber Jensen/Stock, Boston; p.
25—© Liane Enkelis/Stock, Boston; p. 65—© Robert Estall/
Tony Stone World Wide from Click, Chicago; p. 76—
© Frank Cezus/Tony Stone Worldwide from Click, Chicago;
p. 85—© Stephen R. Swinburne/Stock, Boston; p. 93—©
John Coletti/Stock, Boston; p. 94—© D. E. Cox/Tony Stone
World Wide from Click, Chicago; p. 106, 107—© Bob
Daemmrich/Stock, Boston; p. 128—© Paul X. Scott/Sygma

ISBN: 0-8092-4126-9

Published by Contemporary Books,
a division of NTC/Contemporary Publishing Group, Inc.,
4255 West Touhy Avenue,
Lincolnwood (Chicago), Illinois 60712-1975 U.S.A.

0 1 2 3 4 5 6 7 GB(H) 23 22 21 20 19 18 17 16 15 14

Editorial Director
Caren Van Slyke

Editorial
Cathy Hobbins
Ilene Weismehl
Chris Benton
Janice Bryant
Karin Evans
Eunice Hoshizaki

Editorial Assistant
Erica Pochis

Editorial Production Manager
Norma Fioretti

Production Editor
Jean Farley Brown

Production Assistant
Marina Micari

Cover Design
Georgene Sainati

Illustrator
Kathy Dzielak

Art & Production
Carolyn Hopp
Princess Louise El

Typography
Thomas D. Scharf
Ellen M. Yukel

Cover photo © C. C. Cain

Contents

To the Student

Reading to perform specific tasks is very different from the reading done in a classroom at school. Reading in a classroom requires you to learn information and facts to be remembered for future use. Reading to perform tasks, however, involves such tasks as assembling a product or preparing a meal. Usually, in these instances, you read as you perform each step.

Reading Skills That Work presents many tips and strategies to help you improve your ability to "read to do." Not only will you gain practice in reading materials related to everyday tasks, but you will also improve your skill in reading materials that relate to the work world. Some of these tasks include reading to fill out forms; reading to interpret symbols and codes; reading graphs, tables, and charts; and reading to solve problems.

Reading Skills That Work, Book 1 contains 18 lessons divided into six units. Each lesson includes:

- an opening story that presents life situations requiring the use of a particular reading or reasoning skill

- follow-up exercises that allow you to practice the skills and strategies taught in each lesson (these exercises are called *Work Out*)

- *On the Job* activities at the end of each lesson that enable you to apply the skills to work situations

In addition, this book contains:

- unit summaries that recap the key skills taught in each unit

- a final, comprehensive review that gives you a chance to test your mastery of all the skills in this book

- an answer key that explains the answers for the exercises

The activities in *Reading Skills That Work* may be very different from those found in many books you've studied before. Almost all of the exercises are based on written instructions, procedures, forms, diagrams, or memos. Although each *Work Out* exercise has a recommended answer in the key at the back of the book, *finding the answer to each exercise should not be your main goal.* Rather, your goal in using this book should be to master strategies that can help you understand materials you will read in different life and work situations. You will gain more from *Reading Skills That Work* if you focus your attention on understanding the strategies presented in this book instead of just concentrating on getting the right answers to the exercises.

A good way to further sharpen your ability to read for different tasks is to apply the strategies you learn in this book to different situations in your life. Upon finishing the activities in this book, we hope that you will have learned many reading skills and strategies that will work for you throughout your personal and your working life.

Building a Working Vocabulary

Every job has its own working vocabulary.

Most people will play many roles in life—consumer, worker, parent, etc. And, most people will change jobs and companies several times in their lifetimes.

In all parts of life, each place has its own vocabulary. This is especially true in the workplace. Part of a person's success on the job depends on being able to read and use a working vocabulary, a set of special words that describe the job and the workplace. All workers—doctors and nurses, housekeepers and office workers, and others—in a hospital will read words that are special to their job and their workplace. Most of their working vocabulary is different from the vocabulary of workers in a supermarket, a law office, a school, or an auto shop.

Although the words are different from job to job and workplace to workplace, recognizing the words and learning what they mean are the same for every new employee.

The lessons in Unit One can help you build a working vocabulary. You'll learn to read and use different job-related words. You'll also learn some ways to find the meanings for new words.

Recognizing Terms

"*Dredge* meat in flour; then brown over medium heat," Lee read out loud.

"You will dip the pieces of meat in flour so they are coated thoroughly. Then you will fry them until the meat turns a golden brown," Mrs. Factora explained.

"How high is medium on a gas stove?"

Mrs. Factora walked over to their stove. "This is what I do: I turn the heat all the way up to get a high flame. Then I turn it down slowly until the flame becomes half its height."

Lee continued reading the recipe: "*Julienne* the carrots and beans. *Julienne*?"

"You cut them into strips like matchsticks." Mrs. Factora cut a carrot lengthwise. Then, holding her knife, she cut a few strips from one of the pieces.

"The recipe says I have to *mince* onions and slice potatoes. What's the difference between mince and slice?"

"When you slice something, you cut thin, even pieces. But when you mince, you cut very, very tiny pieces like little squares."

"*Julienne, mince, slice* . . . Are there other words cooks use for cutting?"

Mrs. Factora smiled. "You can also

Knowing cooking terms helps cooks do their jobs well.

chop, dice, grate, pare, and *peel.*"

"I wonder if I'll ever understand these cooking words," said Lee. She put the cookbook aside and began peeling a potato.

"When I started to cook, I thought the same way," said Mrs. Factora. "But look at me now!"

Talk about It

- Why do you think cooks would use words such as *mince, slice,* and *grate* instead of just the word *cut*?

- What are other words that are commonly used in cooking?

- Why is it important that you understand all words that are used on a job?

Coming to Terms

Read these sentences about a team. What sport do you think the team plays?

The *team* won its first *game* of the *season*. Both its *defense* and its *offense* were strong.

The words that are in italic are sports **terms**. But they don't give you any clue to the actual sport that the team plays. That's because the words are general sports terms. They can be used in any sport such as baseball, football, volleyball, hockey, or basketball.

Many words that we use in a workplace are general job terms. The words may describe things such as work rules that every workplace has. The terms also may describe the roles that employees fill in a workplace.

▼ Work Out

You'll read these job terms in any workplace. Use them to finish the puzzle.

co-worker   salary   department
policy   memorandum   procedure
employer   supervisor   personnel

Across
4. a person or group that pays others to work for it

6. a person in charge of a group of workers

8. a note that reminds a worker about something important about work or the workplace

9. a person that you work with in a workplace

Down
1. the fixed amount that you are paid regularly, such as once a week or twice a month

2. a section within a company

3. all employees of a company

5. a company rule or guideline

7. the way that a task is done

Special Job Terms

Have you ever noticed how different businesses have their own set of words?

Let's say you go to a store. A salesperson calls you a **customer**. When you see your doctor, she calls you her **patient**. Later you go to a law office. The secretary tells the lawyer that a **client** has arrived for an appointment.

Customers, patients, and clients are people who buy goods or services from a business. Every business will have its own term to describe someone who buys its goods or services. That's because every workplace has its own set of special job terms.

So the words a salesperson uses to talk to you will be different from a doctor's or a lawyer's.

▶ Name one term that each worker might use.

salesperson: _____

lawyer: _____

doctor: _____

▼ Work Out

Find out how vocabulary differs in different workplaces. Choose two different businesses that are near you. Write below the kind of business it is (such as restaurant or law office). Then brainstorm for types of job terms that are used in each workplace. Write five job terms that you read in each workplace.

1. _____
(kind of workplace)

_____

_____

_____

_____

_____

2. _____
(kind of workplace)

_____

_____

_____

_____

_____

What the Terms Mean

As you now know, every workplace has its own
special vocabulary. The job terms will describe people
and materials, tools and equipment, ways that work must
be done, and other work-related things about a workplace.

Special terms describe the equipment these employees use.

▼ Work Out

Match each set of job terms with the workplace where you would hear them. Use
whatever knowledge you already have to help you. The first one has been done for you.

_f_ **1.** social welfare
agency

___ **2.** flower shop

___ **3.** medical clinic

___ **4.** bank

___ **5.** insurance
company

___ **6.** warehouse

___ **7.** restaurant

___ **8.** department store

a. blood pressure, medicine, thermometer

b. beneficiary, premium, policy

c. wreath, arrangement, delivery

d. stock, invoice, inventory

e. discount, receipt, credit card

f. caseload, evaluation, document

g. change, deposit, withdrawal

h. menu, order, entree

Describing an Occupation

Every position in a workplace is given a **job title**. The job title may be made up of two or three words. One word describes the occupation, such as *clerk*. The other words describe the kind of work that the person carries out, such as *clerk typist* and *shipping clerk*.

When you read each word in a job title, you can get a good idea of a position's functions. The job title can give you an idea of the material, people, or objects that a job involves.

The job title for these workers is *shipping clerk*.

▼ Work Out

Read the descriptions for the occupations below. Write the name of each occupation under the appropriate description. The first one has been done for you.

bank teller	**hotel desk clerk**	**salesclerk**
receptionist	**shipping clerk**	**waiter or waitress**

1. signs in guests and assigns them a room

hotel desk clerk

2. takes customers' orders and serves them their meals

_____

3. handles money that customers deposit or withdraw from their accounts

_____

4. packs orders and moves shipments to loading dock

_____

5. greets visitors and answers phones

_____

6. helps customers make purchases

_____

Describing Tasks

▶ Let's say you are preparing a special dinner. Name one task you would need to do to get the dinner ready.

_____

You might have named these tasks: *Plan the food dishes for the dinner. Shop for food. Prepare the food. Cook the food. Set the table.*

Describing tasks that you do at home is the same as describing tasks in the workplace. When you describe a task, you usually state the things or persons that you will work with. And you state what you do with them.

▶ Suppose these sentences describe the tasks for a caterer. Underline the job terms that describe what the cook's helper must *do* in each task. They are *action* words.

> Plan the dishes for the dinner. Shop for food.
> Prepare the food. Cook the food. Set the table.
> Serve the food.

You should have underlined these terms: *plan, shop, prepare, cook, set,* and *serve.*

▼ Work Out

These sentences describe the tasks of a dental assistant. Use the action words to complete them. The first one has been done for you.

Assist	**Operate**	**Perform**
Maintain	**Organize**	**Prepare**

1. P r e p a r e trays and cleaning instruments for dentists.

2. O ___ ___ ___ a ___ ___ x-ray machine and other equipment as needed.

3. M ___ ___ ___ ___ ___ ___ ___ equipment and work areas.

4. O ___ ___ a ___ ___ ___ ___ patients' x-rays and dental records.

5. A ___ ___ ___ ___ ___ dentists in giving presentations on tooth care to patients.

6. P ___ r ___ ___ ___ ___ other tasks as needed.

Knowing Your Job Terms

On any job you'll read and use words that are common to the world of work. Here's a review of the kinds of job terms that you'll read:

- General job terms that are used in any workplace
  Examples: employer, employee, supervisor, department

- Job terms that are special to a workplace
  Examples: The words *blood pressure, medicine,* and *thermometer* are used in a medical setting such as a doctor's office, a hospital, or a medical clinic

- Job terms that describe an occupation and its tasks
  Examples: A dental assistant *preps* the patient and *sterilizes* the cleaning instruments for dentists.

▼ Work Out

In which occupation or workplace might you read or hear each set of terms? Check off each possible answer. A set of words may be used in one or more jobs or workplaces. Use whatever knowledge you already have about these occupations to help you.

1. accounting, invoicing, cash flow
   ____ **a.** bookkeeper
   ____ **b.** accounting clerk
   ____ **c.** editor

2. electrical, wiring, assembly
   ____ **a.** tow-truck driver
   ____ **b.** electronics assembler
   ____ **c.** printer

3. layout, line art, mechanical
   ____ **a.** graphic artist
   ____ **b.** civil engineer
   ____ **c.** social studies teacher

4. salary, vacation, sick leave
   ____ **a.** hotel
   ____ **b.** police station
   ____ **c.** school

5. styling, condition, perms
   ____ **a.** airport
   ____ **b.** chemistry lab
   ____ **c.** hair salon

6. container, waste stream, plastic
   ____ **a.** public bus system
   ____ **b.** recycling center
   ____ **c.** sewing factory

ON THE JOB

What are some job terms that you are familiar with? Answer these questions. Write as many job terms as you can.

1. What general job terms are used in a work setting that you are familiar with?

_____
_____

_____

2. Where would you like to work?

_____

   a. Name some terms that are special to that workplace.

_____

_____

_____

   b. What are the job titles of some of the people who work there?

_____

_____

_____

3. What job title would you like to have?

_____

   a. Name some materials, tools, and equipment that you would work with.

_____

_____

   b. What are some job tasks that you would perform?

_____

_____

_____

Reading "Shorthand"

Scene: KJ and Mel are driving down a freeway. They're going to a barbecue in another city. Mel drives while KJ reads the freeway signs. He's looking for their exit.

KJ: *(spells letters off a freeway sign)* B-L-V-D. What's that mean?

Mel: Boulevard. What boulevard was that?

KJ: Marine World, I think. *(He reads another sign.)* S-T Mary's College. Street Mary's College?

Mel: Saint Mary's College.

KJ: S-T stands for "street" in my book.

Mel: It can also stand for saint. You're getting me nervous, KJ.

KJ: Me! I'm not the one who's driving faster than he should.

Mel: *(He looks at the speedometer. He eases his foot off the gas pedal.)* We are at San Carlos. We should have seen the sign for NASA by now.

KJ: Well, I haven't seen it. Nasa. That's a strange name. What is it—Spanish, Italian, Native American?

Mel: It's an acronym. The letters stand for the first letters in a bunch of words. I forget the name NASA stands for. But

Highway signs make frequent use of abbreviations and acronyms.

you know it. It's the name of the group that sends the astronauts into space.

KJ: The National something Space Administration?

Mel: Something like that.

KJ: You should have told me that NASA was what I'm supposed to be looking for, Mel. I saw the sign ten minutes ago.

Talk about It

- What sign are KJ and Mel looking for?

- Why did they miss their sign?

- What kind of words did KJ have problems reading? Where do you see abbreviations or acronyms? How important is it to be able to read those kinds of words?

Reading Abbreviations

What words can you read in the job advertisement at the right?

To save space, newspapers will often use **abbreviations** in their advertisements. The words are made shorter by taking certain letters out. The abbreviation **wrkrs**, for instance, stands for the word **workers**.

What do you think the abbreviation **FT** is for: **fun time**, **full-time**, or **fast track**?

You'll often read abbreviations in the workplace. Some abbreviations will stand for words that name people, places, and things. Others will stand for units of weight and measure.

Examples: **pres.** means **president**; **FL** means **Florida** **pg.** means **page**; **lb.** means **pound**

To read an abbreviation, you say the word. So when you see the abbreviation **lb.**, you say the word **pound**. Here are some abbreviations you already know. Read them aloud: **Mon., Tues., Wed., Thurs., Fri.**

▶ Now look at the job advertisement at the top of this page. Try to figure out what all the abbreviations stand for. Write them here:

_____

_____

You should have written: *part-time, full-time, workers wanted, good hourly wage, good benefits,* and *equal opportunity employer.*

| PT, FT wrkrs wtd. |
| Gd. hrly. wg. |
| Gd. ben. EOE. |

▼ Work Out

Read each sentence. Write the abbreviation for the term that is in **bold** type. Use the abbreviations in the box to help you. The first one has been done for you.

tsp. 1. The recipe calls for one **teaspoon** of nutmeg.

____ 2. Please complete the form **as soon as possible**.

____ 3. **Females** are encouraged to apply for the manager's job.

____ 4. Take out all invoices from the **miscellaneous** file.

____ 5. This memo is in **reference** to the staff meetings.

Abbreviations	
tsp.	ASAP
F	ref.
misc.	Sept.

Abbreviations for Phrases and Long Names

Sometimes you'll read an abbreviation that stands for a phrase or a proper noun. The letters in the abbreviation are the first letters of each word.

Examples: **TGIF** for **T**hank **G**od (or **G**oodness) **I**t's
                 Friday
              PTA for **P**arent-**T**eacher **A**ssociation
              IBM for **I**nternational **B**usiness **M**achines

Think of an abbreviation for a phrase or name that you know. What does the abbreviation stand for?
You can read abbreviations for phrases or names in one of two ways. You say the complete phrase or name. Or you say each of the letters in the abbreviation. For instance, when you read **TGIF**, you say "Thank God (or Goodness) It's Friday" or "T-G-I-F."

If the letters are generally spoken as a word, they form an acronym, a special kind of abbreviation. The word *scuba*, **s**elf-**c**ontained **u**nderwater **b**reathing **a**pparatus, is an example of an acronym.

▼ Work Out

Here are some abbreviations that you might read in newspapers and in the workplace. Match them with their full names and phrases. The first one has been done for you.

_j_ **1.** COD	**a.** Federal Standards Labor Act
___ **2.** SSA	**b.** National Labor Relations Board
___ **3.** FYI	**c.** Equal Employment Opportunity Commission
___ **4.** FSLA	**d.** Internal Revenue Service
___ **5.** EOE	**e.** Social Security Administration
___ **6.** VIP	**f.** standard operating procedure
___ **7.** NLRB	**g.** for your information
___ **8.** IRS	**h.** very important person
___ **9.** EEOC	**i.** equal opportunity employer
___ **10.** SOP	**j.** cash on delivery

Reading Acronyms

Acronyms are a kind of abbreviation for fast communication. Acronyms are words. You'll sometimes read them in the workplace. The letters in an acronym stand for the first letters or syllables in a name or a phrase.

acronym: a word that is formed from the letters or syllables of a series of words

For example: **NASA** means **N**ational **A**eronautics and **S**pace **A**dministration; **radar** means **ra**dio **d**etecting **and r**anging

People often recognize an acronym more easily than the full name of a group or thing.

▶ One acronym that we often read or hear is **z**one improvement **p**lan code. What is its acronym?

_____

Its acronym is *zip code.*

▼ Work Out

Here are some acronyms that you might read in a workplace. Match the acronym with the full name. The first one has been done for you.

_d_ **1.** HUD	**a.** Occupational Safety and Health Administration
____ **2.** OSHA	**b.** Wide-Area Telecommunications Service line
____ **3.** ESOP	**c.** Federal Insurance Contributions Act
____ **4.** IRA	**d.** [Department of] Housing and Urban Development
____ **5.** WATS line	**e.** Individual Retirement Account
____ **6.** FICA	**f.** Employee Stock Ownership Plan
____ **7.** AIDS	**g.** Organization of Petroleum Exporting Countries
____ **8.** PAC	**h.** Disk Operating System
____ **9.** DOS	**i.** United Nations International Children's Emergency Fund
____ **10.** UNICEF	
____ **11.** OPEC	**j.** Special Weapons and Tactics
____ **12.** SWAT	**k.** Acquired Immunodeficiency Syndrome
	l. Political Action Committee

Reading Symbols

Here's another kind of "shorthand" term that you'll read in the workplace: symbols. They may be numbers, letters, shapes, colors, or drawings. Some symbols are like abbreviations. They stand for words and phrases. For example, the symbol & (ampersand) stands for "and."

Symbols also stand for ideas. We read that kind of symbol when we drive. For example, here's one:

symbol: a figure, shape, or other thing that stands for a word or an idea

⬡ The word and the shape and color of the sign make up the symbol.

▶ What idea does that symbol stand for?

_____

The *stop sign*, like many symbols that we use, is standard. The meaning of the symbol is the same almost everywhere you go. So reading a stop sign in a factory in Texas tells us the same thing as reading one near a cane field in Hawaii.

▼ Work Out

You'll read these road signs anywhere in the United States. Match each symbol with its meaning. The first one has been done for you.

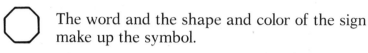

_e_ 1. 🚸   **a.** The exit for the airport is ahead.

___ 2. 🚫   **b.** You can get gasoline at the next exit.

___ 3. ✈   **c.** Here's a public pay phone.

___ 4. ⊗   **d.** A hospital is nearby.

___ 5. ⛽   **e.** Watch out for students going to and from school.

___ 6. 🅟   **f.** A railroad track is coming up. Watch out for trains.

___ 7. **H**   **g.** Do not park in this place.

___ 8. ☏   **h.** Do not make a left turn.

Reading Symbols to Operate Machines

Many machines in the workplace will have symbols on them. You read the symbols to help you operate, or work, a machine. For instance, look at the picture of a calculator. Some buttons have number symbols. Others have math signs.

To operate the calculator, you press the button with the correct symbol. Let's say you're ready to get a total for a problem. Which button will you press? (Ask yourself: Which math sign means "equal"?)

▶ Now let's say you need to do each of these math operations. What buttons will you press? Write the correct symbol.

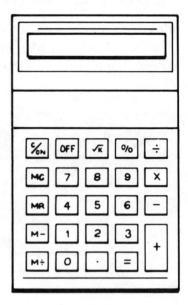

Add _____   Subtract _____

Multiply _____   Divide _____

▼ Work Out

Here are some symbols that you would find on a copy machine. Answer the questions by writing the letter that is over the symbol. The first one has been done for you.

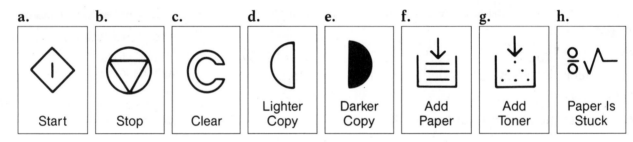

a.	b.	c.	d.	e.	f.	g.	h.
Start	Stop	Clear	Lighter Copy	Darker Copy	Add Paper	Add Toner	Paper Is Stuck

1. You press a button to clear the machine of commands the last person gave it. What symbol is on the button?

_c_   _c_

2. You press a button to make your letter come out darker. What symbol is on the button?

_____

3. You're ready for the machine to copy your letter. What symbol is on the button that you press?

_____

4. You want the machine to stop copying your letter. What symbol is on the button that you press?

_____

5. Paper gets stuck in the machine. A red light comes on, and a certain symbol flashes. What symbol is it?

_____

6. The machine signals you that paper needs to be added. What symbol would be flashing?

_____

Reading Codes

Codes are another kind of shorthand used in the workplace. A set of codes may be numbers, letters, or symbols. Each code will stand for a certain fact about a subject.

These postal workers sort mail by zip codes.

Your zip code is part of a system of codes. The U.S. Post Office assigns zip codes to every place in the United States. When postal workers read a zip code, they know where to deliver the mail.

Sometimes a set of codes like the one at the right appears on paychecks.

The codes stand for amounts that might be deducted from a paycheck. Let's say $24 is deducted. The code **4** was next to it.

▶ What is the amount subtracted for?

_____

▶ Suppose $14.50 is deducted for a code **2**. What is the amount subtracted for?

_____

▶ A paycheck shows $8.25 deducted for a code **5** and $15.00 for a code **6**. What is each amount subtracted for?

_____

You should have written these answers: Code **4** stands for union dues. Code **2** stands for health insurance; Code **5** for retirement plan; and Code **6** other.

▼ **Other Deductions**

1 thrift and stock
2 health insurance
3 United Way
4 union dues
5 retirement plan
6 other

▼ Work Out

You might read passages like this in job manuals. Read the passage and answer the questions.

You might read this in a secretary's manual.

Memorandum Format

A memorandum must have the following headings printed at the top of the page: "TO," "FROM," "DATE," and "SUBJECT."

On the "TO" line, type the name of the person or persons who will receive the memorandum. Type the name of the writer on the "FROM" line. Type the date that the message is written on the "DATE" line and the subject of the message on the "SUBJECT" line. The writer of the memorandum must sign his/her initials near the name.

---

1. What is the subject?

_____

2. What is the main point in the first paragraph?

_____

_____

3. What is the main point in the second paragraph?

_____

_____

4. Check off the sentence that best states the main point of the text.

_____ Certain information must be the same on every memorandum.

_____ All memorandums must have the name of the writer.

_____ Memorandums are sent between co-workers in a company.

▼ Work Out

You might find this information in an employee's handbook. Read the text and answer the questions.

Your Employee's Handbook

1 This handbook explains the company's policies and programs. By reading each section, you have an overview of the work rules as well as the benefits that you receive as an employee of the company. This handbook is not an employment contract.

2 We have the right to change any of the policies or programs that are presented in this book. You will be given a description of any changes that we make.

3 All employees are given a handbook. You will sign a receipt to show that you have been given a copy. When you leave the company, return your copy to your supervisor or the Personnel Department.

---

1. What is the subject? _____

2. What is the main point of each paragraph?

Paragraph 1: _____

_____

Paragraph 2: _____

_____

_____

Paragraph 3: _____

_____

3. What is the main point of the whole text?

_____

_____

ON THE JOB

Imagine that you're a nurse assigned to a day-care center. Your supervisor tells you to read this passage in a manual. Read it; then answer the questions.

Giving Ear Drops

Make sure you have the right drops. Get the bottle of ear drops ready for the child. Read the prescription on the bottle. Be sure you know how many drops you must give the child. Warm the bottle by setting it in a bowl of warm water for a few minutes.

Give the child the drops in the first-aid room. Have the child lie on his or her side. Let the drops fall gently into the center of the ear. Count the number of drops as you put them in. The dropper should not touch the child's ear. If it does, wash the dropper before putting it back in the bottle.

---

1. What is the subject of the passage?

2. What is the main point of the text?

3. What is the main point of each paragraph in the text?

Paragraph 1: _____

Paragraph 2: _____

4. What are the specific details that support the main point about the subject? Make a list of them. Write the details in your own words. Number the details **1**, **2**, **3**, and so on.

Paragraph 1	Paragraph 2

Understanding the Message

Here are some things you can do to help you understand signs and labels:

- Look up the meaning of any terms that you don't know.

- Name the subject of the sign or label.

- Determine the main point. Ask yourself: What *one* big idea do the details support?

▼ Work Out

Here's a work rule that you might read in the workplace. Answer the questions about the message.

If you must smoke, please smoke *only* in allowed places

1. What is the subject of the sign?

_____

2. What details are given?

_____

_____

_____

_____

_____

3. What is the main point of the message?

_____

_____

ON THE JOB

Imagine that you work in a chemical lab. You work around many dangerous chemicals. One of the safety rules you must follow is this: **Wear gloves whenever you handle chemicals**.

Make a sign that brings out the message clearly. You can use text, pictures, or both to show your details and make your point. Use the space below.

UNIT TWO SUMMARY

In the workplace you'll read materials such as memorandums, forms, manuals, labels, and signs. To interpret the written ideas, you need to understand the details and main point.

To help you find the details and main points of work materials, remember:

- Identify the subject of the text or message.

- Identify the kinds of details that are given in a paragraph.

- Identify the *single* main idea that all the details support.

UNIT CHECK

Answer the questions.

1. What are three things that you can do to understand any memo?

    a. _____

    b. _____

    c. _____

2. Why is it important to identify the main point of a text or sign?

_____

_____

TALK ABOUT IT

1. Discuss work habits and attitudes that are important to have for any job. Write a paragraph about one habit or attitude.

2. List as many signs and labels as you can find in your workplace or that you remember from a place where you once worked.

3. If you are employed, copy or make a sketch of a sign from your workplace. Discuss: Why is the message important? Is the main point clear from reading the sign? If it's not clear, how could you fix the sign so that the message is clearer?

FOR YOUR INFORMATION

At many vocational training programs you can get manuals like the ones that workers use on their jobs. Look up an occupation, such as nurse's aide, that you may be interested in. Or look up a subject, such as plumbing or computer programming.

Making Graphics Work for You

A graphic effectively gets across this supervisor's point about workplace safety.

Graphics are pictures, maps, charts, or graphs that are used to illustrate information. People respond to pictures more readily than they do to words, so graphics are being used more and more every day.

Many of us see graphics so often that we ignore them when we see them. Yet graphics can help us understand the main point of something. In fact they can often explain the main point more clearly than words.

On the job, graphics can be helpful tools. Some graphics can show you how things work. Some graphics can remind you of what steps to take in a job. Other graphics can help you solve problems. And still other graphics can help motivate you to keep doing your job.

In Unit Three you'll learn about several kinds of graphics. You'll learn to make sense out of them so that you can make them work for you.

A Picture's Worth a Thousand Words

"What a place to have a flat tire," said Bobby. He eyed the cars racing down the freeway.

Laura gave her son the car manual. "Find the pictures that show how to change a flat," she said. She opened the trunk and began to unscrew the bolt securing the spare tire.

"We're OK, Bobby. I've done this before," said Laura. "I just need the diagrams to remind me what to do."

Cars whizzed by, shaking the disabled *Escort*.

Bobby flipped through the manual, found the diagrams his mother wanted, and held the manual open for her to study.

Laura read quickly. She picked up the lug wrench. "Bobby, take the jack and this long tool. When I've loosened the nuts, you'll jack up the car."

"Gotcha," Bobby replied.

Laura showed him where to set the jack and what he would have to do. Then she went to work: She pulled off the hubcap easily. But the nuts were so tight that she had to work hard to turn the wrench. What she had expected to be a snap was taking much longer. After loosening the fourth nut, she looked at her son. "This is kind of fun," she said.

"Mom, you need to get out more if you call this fun," Bobby said, grinning. He was proud of her. This was the first time he had ever seen his mother change a flat.

"There, the last one's loose. OK, Bobby, raise her up."

Bobby jacked up the car until the tire was off the ground. Together the mother and son took off the nuts and pulled off the tire. Together they put the spare tire in its place and screwed the nuts back on. They worked quickly.

As Bobby lowered the car to the ground, a tow truck slowed down and pulled up. A tall, stocky man got out of the truck. "Looks like you could use a hand. Can I help?"

Laura smiled. She put her arm around Bobby. "Thanks, but we've already finished the job."

They got into the car and Laura stepped on the accelerator, easing the *Escort* into the flow of traffic.

"Good job, Mom," Bobby said.

"Now that you know how, next time you can do it yourself," Laura said smiling.

Talk about It

- What diagrams do you think Laura studied?

- Why do you think Laura wanted to look at pictures rather than read words?

- When have you used a diagram to fix or put something together?

Reading Pictures for Details

Here is a text that you might read in a car manual. It describes how to jack up a car.

Insert jack handle into jack. Turn the jack handle **clockwise** to raise the jack.

If you work with cars or have fixed a flat before, the text should have been easy to understand. But if not, you probably didn't understand it at all.

Now look at the **diagram** below. It shows what the words are describing. Do the words make more sense now?

Diagrams can help readers understand ideas. They show what things or parts of a thing look like. Labels on a diagram identify what the things or parts are.

By studying the pictures and reading the labels, you can see how unfamiliar things work.

clockwise: in the direction that the hands of a clock move.

diagram: a graphic that explains and shows the relation of parts to each other.

▼ Work Out

Answer the questions about the diagram.

1. Describe the appearance of the jack handle.

   _____

2. When you turn the jack handle counterclockwise, what happens to the jack?

   _____

3. Suppose you have finished changing a flat. How would you lower the car?

   _____

4. In an emergency such as a flat tire, why is a diagram more effective than a paragraph of directions?

   _____

   _____

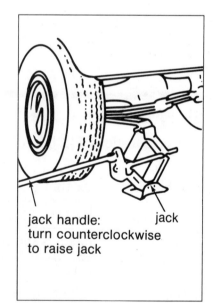

jack handle: turn counterclockwise to raise jack

jack

Seeing Connections

Some diagrams show how **systems** work. A system is made up of several parts. A diagram shows how the parts are connected to each other so that the whole system can work.

system: a group of items that work together to perform a function

For example, the diagram on this page shows a telephone answering machine system. The system has six parts.

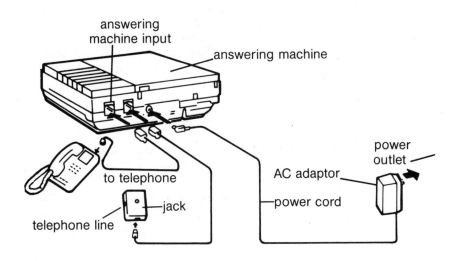

▼ Work Out

Answer questions 1–3 based on the diagram above.

1. What is the part that goes to the power outlet called?

_____

2. What part or parts is the phone directly connected to?

_____

3. Suppose you pulled out the power cord. Do you think the phone would still work? Why?

_____

_____

4. Suppose you turned off the answering machine. Do you think the phone would still work? Why?

_____

_____

Following Diagrams in the Workplace

When you read diagrams in the workplace, they come with written directions. Study a diagram carefully and read everything that is on it: titles, captions, and labels. To understand a diagram, be sure you know its subject. And be sure you know its main point. Remember: find the chief idea that the details of the diagram describe.

▼ Work Out

Answer the following questions.

1. What is the subject of the diagram?

   _____

2. What is the main point of the diagram?

   _____

3. What part shows how much substance is inside a fire extinguisher?

   _____

4. What part would you aim at a fire?

   _____

5. What three things should you do when you use a fire extinguisher? Check each one.

   _____ **a.** Check pressure gauge.

   _____ **b.** Remove pull pin.

   _____ **c.** Take hose off the hose clip.

   _____ **d.** Clean off nozzle.

   _____ **e.** Press down on lever.

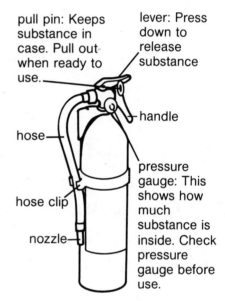

pull pin: Keeps substance in case. Pull out when ready to use.

lever: Press down to release substance

hose

handle

hose clip

pressure gauge: This shows how much substance is inside. Check pressure gauge before use.

nozzle

Parts of a Fire Extinguisher

ON THE JOB

Imagine that you work in a small office. Your office has just bought a personal computer. Your supervisor will teach you how to use it. She has given you this diagram to study. Read it; then answer the questions.

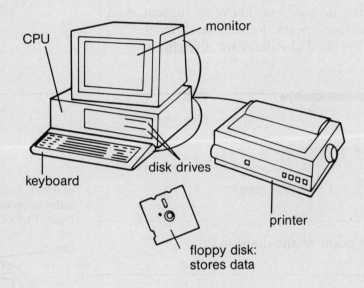

1. What five parts make up the computer system?

   _____

2. Where is the information that you put into the computer stored?

   _____

3. What is the function of the monitor (screen)?

   _____

4. What is the function of the printer?

   _____

5. Is the printer connected to the keyboard or the CPU (Central Processing Unit)?

   _____

Finding Your Way

Gilda Scott stepped out of the hospital elevator, trying to remember the directions the information clerk had given her. Go to the right. Turn left at the corridor. Halfway down is the medical clinic. The eye clinic is across from it.

She didn't find any corridor. Instead she found the skywalk that led to the next building. I must have heard her directions wrong. Maybe I was supposed to go left, she thought.

She walked back to the elevator. She walked past it and turned right into another corridor. She came to the children's clinic.

Gilda moaned. She hated being lost. She walked up to the counter at the children's clinic. A nurse was seated behind the counter filling out a patient's chart. "Excuse me. Can you tell me where the medical clinic is?"

The nurse smiled. Gilda could tell that the nurse was used to directing visitors who got lost. "First, you're in the wrong building. But that's OK. You're on the right floor. Maybe this map will help you get your bearings."

She pulled out a floor map and put it in front of Gilda. She put an X on the map. She said, "You're *here*. Now go back to the elevators. But don't get on. Walk to the end of the corridor. It will take you to the skywalk. The skywalk connects the two buildings.

"Cross the skywalk to the other building. Turn right. You'll pass more elevators. Turn left at the first corridor you come to. About halfway down that corridor is the medical clinic. It's on the left side."

The nurse traced the route on the map as she spoke. She handed it to Gilda. "Just follow the red line on the map. You can't go wrong."

"Thank you," Gilda said. She turned away from the counter. She looked gratefully at the map. She couldn't remember all of the nurse's directions. But she knew that the map would get her there.

"Thank goodness for maps," she said aloud.

Talk about It

- What does the nurse use to show Gilda where to go?

- Why does Gilda seem glad that she has a floor plan to help her get to the medical clinic?

- Why do you think that city ordinances often require public buildings to post floor plans?

- What other things could floor plans and maps help you with?

Reading a Map

In Lesson 7 you learned about graphics that can show you how something works. Other kinds of graphics can help you decide where to go or what to do.

You probably think of maps that show a country or the streets of a town. But maps are also made to show the layout of a room, floor, building, or set of buildings such as a school campus.

The map on this page shows the layout of a floor in a hospital.

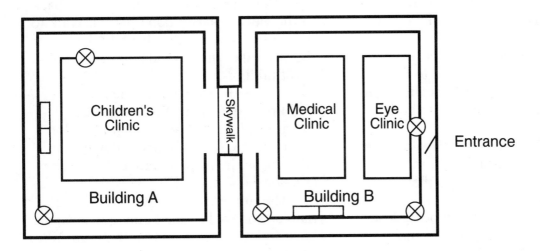

Look at the map. Notice that it shows two buildings. A skywalk connects the two buildings.

What clinic is in Building A? (The *Children's Clinic*)

What clinics are in Building B? (The *Medical* and *Eye Clinics*)

Maps use symbols and labels to show where things are. The symbol ⬚ stands for elevators. Each building has a set of two elevators. Find each set and write *elevator* next to it.

The symbol ⊗ stands for *exit*. Building B has three exits. See where the exits are and circle them. Building A has two exits. Find the exits and write *exit* next to each one.

You read on the previous page about someone who wanted to get to the medical clinic from the children's clinic. This map could help her find her way around the complex.

Getting Around with a Map

To use a map, you must understand the symbols. Many maps have a **key** that explains the symbols.

To use a map, first find the place where you are. Next find the place where you need to go. Now, with your finger, trace a path between the two places. Imagine yourself walking the path. Say where you must turn left or right or go straight ahead.

key: an explanation of the symbols found on a map or chart

▼ Work Out

The map at the right shows the first floor of a hospital. Answer the questions about it.

1. Suppose you get off the elevator near the front door. You must go to the pharmacy. How would you go?

   _____

   _____

2. Suppose you enter the hospital through the back door. You must go to the check-in counter. How would you go?

   _____

   _____

3. Let's say you're waiting for your glasses to be made. You go to the gift shop. How do you get back to the optical lab?

   _____

   _____

4. Suppose you're a clerk in the women's clinic. You must deliver a file to the second floor. You decide to take the stairs by the optical lab. How would you get to the stairs?

   _____

   _____

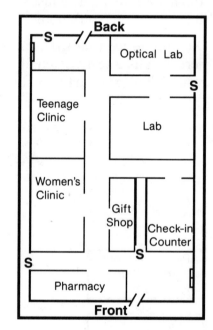

Key

// Entrance/Exit
|| Corridor
▭ Elevator
S Stairs

Finding Out What to Do

You've seen how a graphic can help you get somewhere. Another kind of graphic can help you follow a process. This graphic is called a **flowchart**.

A flowchart is easy to read. Different symbols stand for the steps that must be taken. Arrows pointing to and from the symbols show the order in which the steps must be taken.

The flowchart below shows the steps to get a driver's license. Workers at the motor vehicle department might use it. People who want to get a license might use it as well.

The first step in the procedure is completing an application form. The last step is sending a permanent license to a person's home.

Did you notice that the chart shows what happens if a person fails a test? Often, if a step can have more than one result, the flowchart will show the choices.

flowchart: a diagram that shows a step-by-step process using arrows and symbols

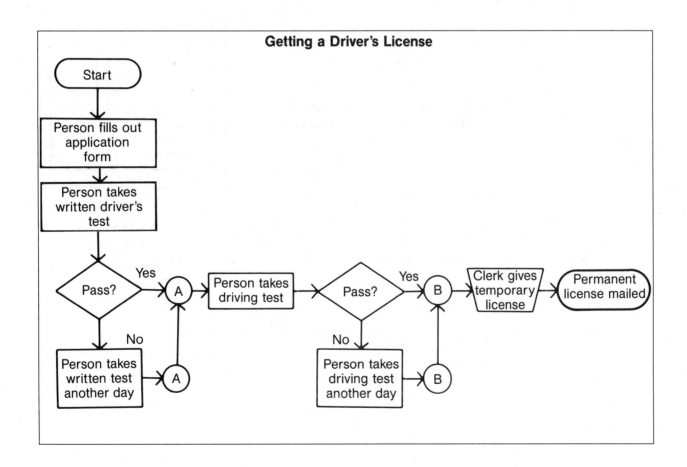

Getting a Driver's License

Imagine that you're a clerk at a motor vehicle department. You're given the flowchart on page 58 to remind you of the department's procedure for giving a driver's license. Study the chart and answer these questions:

1. The flowchart shows five different symbols that are part of the system. These symbols are standard and appear on all flowcharts. They include: a rectangle, diamond, circle, trapezoid (▽) and an oval.

   a. Draw the symbol that stands for a process or step to be performed.

   b. Draw the symbol that shows where the process begins and ends.

   c. Draw the symbol that tells the reader to go back to the first point named by it.

   d. Draw the symbol that always asks for a decision.

   e. Draw the symbol that stands for output—something that the system puts out.

2. If a person successfully completes each step of the procedure, how many steps result in a license being issued? _____

3. If a person passes the written test but fails the driver's test, does he or she have to fill out another form? _____

4. A person who passes the written test but fails the driving test the first time around has completed how many steps by the time he or she gets a permanent license? _____

5. The symbol B in this flowchart tells the applicant who passes the driving test the second time around to go back to the first point named B. What happens at the first point named B?

   _____

ON THE JOB

Suppose you're the mail clerk for the Noel Company. You must follow certain procedures for delivering the mail. Study the flowchart and answer the questions that follow.

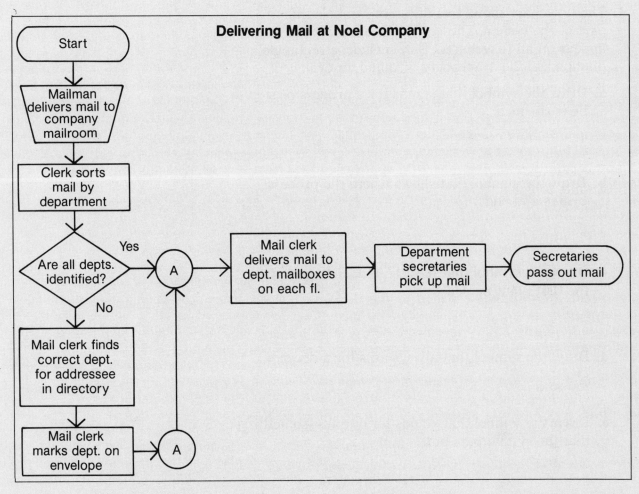

Delivering Mail at Noel Company

1. What is the purpose of the flowchart?

   _____

   _____

2. What question must the mail clerk answer before he can deliver the mail?

   _____

3. How many steps must the mail clerk perform if all of the departments are clearly identified on the envelope? _____

4. How many steps must the mail clerk perform if all of the departments are *not* clearly identified on the envelope or package? _____

5. What happens first: the sorting of mail by department or by employees in a department? _____

   _____

6. Who sorts mail by employees' names in a department? _____

7. What do you think the mail clerk should do if mail arrives for an employee who no longer works for Noel Company and whose former department is *not* identified?

   _____

   _____

Getting Help from a Table

Josie was on the third floor of Daybrook's Department Store.

She headed for the display of bright red, yellow, and blue baby clothes. She broke into a big smile. They were cute. She looked at a price tag. And they were inexpensive.

"May I help you?" a salesclerk asked.

"These are cute," said Josie. "I've never seen anything like them before."

The salesclerk smiled. "Aren't they? They just came in. I just love them. So cheery-looking. Easy to wash. And the price is just what I can afford. They're attracting quite a bit of customer traffic."

"You said it," said Josie. "I can't believe the price of baby clothes these days. Twenty dollars for a baby's top that would last maybe one month."

"Is this your first child?"

"My fourth. It was a surprise. My others are almost teenagers. I gave away all their baby clothes, so I'm starting all over again."

The salesclerk picked up a sleeper and turned a tag over. She showed Josie a size chart. "These clothes are made in Sri Lanka. So their sizes are different from what we're used to."

"Where in the world is Sri Lanka?" Josie asked.

"It's located off the southern coast of India," the clerk replied. "I hadn't heard of the place myself, but since I've begun to get so many questions from customers, I thought I had better find out."

"Thanks." Josie eyed the size chart. "With these prices I could get a few of each size."

"I'll leave you to your fun," said the salesclerk. "If you need help, just ask."

Josie nodded. She was already setting aside things that she might buy.

Talk about It

- What graphic does the salesclerk show Josie?

- What kind of information does it show?

- Why would Josie need to refer to a table?

- Have you ever used a table of information when you shopped?

- Have you ever used a table of information at work?

▼ Work Out

Here's a table that a nurse might use. Read the table. Then answer the questions.

Immunizations for Children							
Immunization	**Age (months)**						
	2	4	6	12	15	18	60 (5 years)
Diphtheria	x	x	x			x	x
Whooping cough	x	x	x				
Tetanus	x	x	x			x	x
Polio	x	x	x			x	x
Measles					x		
German Measles					x		
Tuberculosis (TB) Test				x			

1. What is the subject of the table?

   _____

2. What is the main point of the table?

   _____

3. A parent brings in her six-month-old child for his immunization. What immunizations will the nurse give him?

   _____

4. A parent brings in her fifteen-month-old child for two immunizations. What are they?

   _____

5. A parent has one-year-old twins. She wants to know what immunizations they will need now and in the coming year. What are they?

   _____

6. A child will be turning five years old. What immunizations does a nurse tell his parent that the child will need?

   _____

ON THE JOB

A gardener's helper might read a troubleshooting table like the one below. Study the table; then answer the questions.

Problem	Solution
Soil is too acidic.	Add lime to soil.
Lack of nutrients—nitrogen, phosphorus, and/or potassium—in soil.	Treat soil with the right kind and right amount of fertilizers.
Too much salt in soil.	Now and then water soil thoroughly to wash salt out.
Layer of soil near surface is so hard that plants can't root and water doesn't drain. (This layer is called *hardpan*.)	Plow soil at least 12 inches deep.

1. Why would you use this table on your job? _____

_____

2. A client complains that all the plants you put in several months ago are dead. You look at his garden. You see deposits of salt on the soil. What advice do you give him?

_____

_____

3. You're planning a lawn for a client. You test the soil. It is hardpan. What problem will that cause the lawn?

_____

_____

4. A client tells you that everything she plants dies. You test the soil in her garden and tell her to add lime to the soil. Why?

_____

_____

5. A client's soil doesn't have enough nitrogen. How will you solve the problem?

_____

_____

Seeing the Changes over Time

Scene: Lily and Evie, both in their 40s, are at the table in Lily's kitchen.

Evie: Lily, this fried chicken is delicious. I hope I'm not ruining your diet.

Lily: (laughs) I didn't *stop* eating my favorite foods. I just eat *less* of them.

Evie: (surprised) And you still lost forty pounds! I don't think I could have done it. What did you do? Did you fast?

Lily: I used good old-fashioned willpower. Daydreaming about how I want to look also helped. But a lot of it was . . . I'll show you.
   Lily leaves the room. She returns with a large binder. She puts it in front of her sister. Evie flips through the binder.

Evie: Food articles . . . calorie charts . . . menu substitutes . . . I don't get it. What does all of this have to do with losing forty pounds?

Lily: That's my "Fat Binder." I have articles about nutrition and calorie charts—even for junk food. And I found tables that show how many calories you can lose by performing different activities. But the best is this line graph.

Every Saturday night I put down how much weight I lost that week. Seeing the line move down helped motivate me.

Evie: I remember how you hated school, especially math and reading. Now look at you. Nobody could have convinced me that you'd be graphing your weight.

Lily: Sometimes I can't believe it myself. But I really enjoy all this collecting and reading and sorting and even making my own tables.

Evie: I'm proud of you, Lily. What willpower!

Lily: (She walks over to a counter and picks up a binder) Since I no longer buy so much food, I've saved a lot of money. So I've decided to take a trip to reward myself. This is the best part of the entire program.

Evie: That's great, Lily. Where are you going?

Lily: (She opens the binder to the last page) Hawaii. (Lily gets up and performs a hula dance for Evie.)

Talk about It

- What kind of graphics does Lily have in her "Fat Binder"?

- How did the line graph help Lily lose weight?

- Why do you think graphics such as a line graph could help motivate people to lose weight, earn more money, or do other things?

Using a Line Graph

Line graphs let you see how something has changed over time. The line graph that you'll study in this lesson allows you to see a **trend**. That is, it allows you to see how one amount changes over a period of time.

trend: a line of general direction or movement

For example, you read on the last page about a person who lost weight. The line graph below shows her weight loss for the first ten weeks.

Each point stands for her weight at the end of a certain week. The line allows you to see the ups and downs of the weight loss more easily. By studying the line, you get an idea of how successfully the person is losing weight.

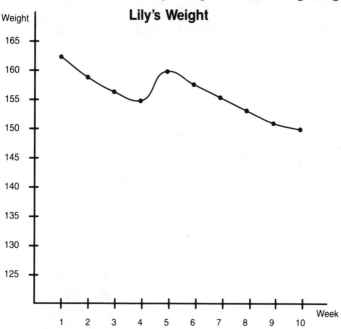

Weight — Lily's Weight — Week

▼ Work Out

You may see line graphs like the one above at any of these places. Match each place with the information that might show on a line graph.

_____ **1.** a bank

_____ **2.** a health clinic

_____ **3.** a library

_____ **4.** a social service agency

_____ **5.** a state employment office

a. how many books were borrowed each month in 1990

b. how many children were immunized against measles each month in 1990

c. what the interest rate was each day during the month of June

d. how many people were out of a job for the last six months of 1989

e. how many clients were seen by caseworkers each month in 1990

Reading a Line Graph

Line graphs are easy to read. Here's what you need to remember: a line graph has two axes—a **horizontal axis** and a **vertical axis**. Look at the line graph below to see what each axis looks like.

The axis that goes from left to right usually shows the units of time. The units may be hours or days, months or years, and so on. Each mark on the scale stands for one unit of time.

The axis that goes up and down shows the amount that is being measured. The amount may be in dollars, pounds, or another kind of unit. Each mark stands for one unit.

horizontal axis: the line running left to right that shows values along a graph

vertical axis: the line running up and down that shows values along a graph

The line graph at the right shows an education budget for the years 1986 through 1990.

▶ Look at the horizontal axis. What unit does each mark stand for?

_____

▶ Now look at the vertical axis. What unit does each mark stand for?

_____

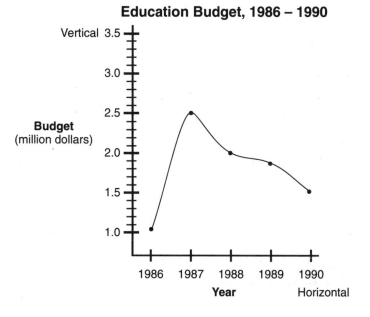

Each mark on the time axis stand for one year. Each mark on the dollar axis stands for a certain money amount: one-tenth of $1 million ($100,000).

Each point on a line graph stands for an amount at a certain time. To read the amount, you read the mark on the dollar axis that's directly across from the point. Then you read the mark on the time axis that's directly beneath the point.

▶ Look at the point that is farthest to the left. What amount does it stand for?

_____

You would read the amount as $1.0 million in 1986.

Practice reading aloud the other points on the line graph.

▼ Work Out

You may see line graphs like the one above at any of these places. Match each place with the information that might show on a line graph.

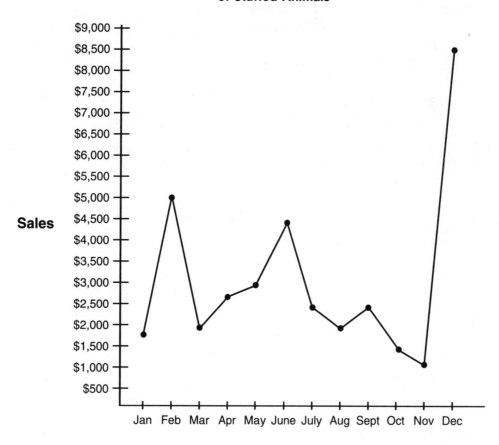

**1990 Monthly Sales
of Stuffed Animals**

1. What is the subject of the line graph?

   _____

2. What is the main point?

   _____

3. In what month did the store make the fewest sales?

   _____

4. In what month did the store make the greatest sales?

   _____

5. What interesting conclusion can you draw about the great rise in stuffed animal sales in December?

   _____

   _____

Comparing Two Trends

Sometimes a line graph is used to show two or more different trends. These line graphs allow you to compare the changes in the trends during the same time period.

For example, a computer store may want to see how many of two different brands of computers it sold from January through June. The line graph below shows the results.

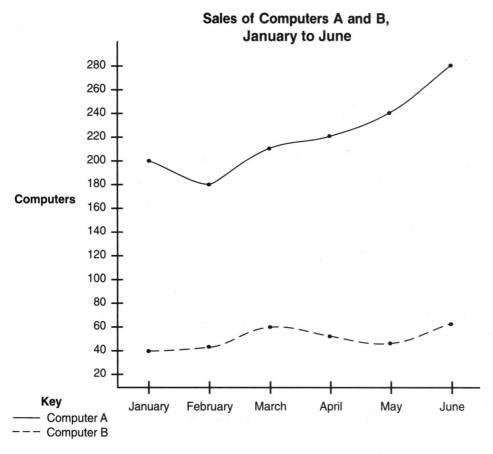

Sales of Computers A and B, January to June

Key
—— Computer A
– – – Computer B

▶ Look at the line graph. How do you know which line stands for computer A?

_____

You probably looked at the key to decide which computer a line stands for. The line for computer A is solid; the line for computer B is broken.

Some line graphs identify a line with a label. Other line graphs use a key. The line graph above uses a key. Read the key; then identify each line by writing *computer A* or *computer B* next to it.

Here are some things you can learn from line graphs that show different trends:

1. You can see how one trend differs from another.

2. You can see in which months results are better than others.

▶ Which computer brand does the store sell more of?

_____

▶ In which month were sales for computer A the highest? For computer B?

_____

You should have written these answers: *The computer store sells more of computer A than computer B, and sales were highest for both computers in June.*

▶ Let's say you had to compare the monthly sales of four computer brands for one year. Which graphic would you prefer to use to show the sales information—a table or a line graph? Why?

_____

_____

A line graph is the best tool to use to show sales trends over periods of time. The visual rising and falling of a line makes it easy to see trends.

▼ Work Out

Answer the questions about the line graph on page 72.

1. How would you describe the sales trend for computer A from January through June?

_____

2. How would you describe the sales trend for computer B for the same time period?

_____

3. How are the two sales trends the same?

_____

4. Describe how the two sales trends are different.

_____

ON THE JOB

Let's say you're a nurse's aide. You're given the line graph below to study. Study the line graph; then answer the questions.

The graph shows the average temperatures a person may have after being exposed to the flu virus or while recovering from hypothermia (below-normal body temperature caused by prolonged exposure to the cold). The normal body temperature is 98.6 degrees Fahrenheit.

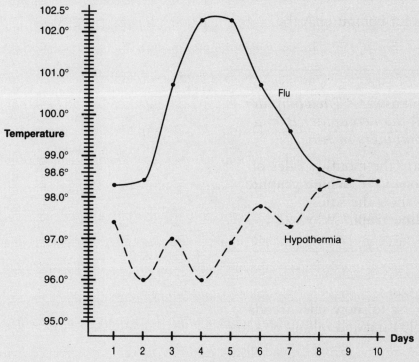

1. What do the marks stand for on the time scale?

_____

2. What amount does each long mark along the temperature scale stand for?

_____

3. What illness, shown on the graph, does a person have who is:

   a. regaining body heat?

   _____

   b. hot from too much body heat?

   _____

4. How would you describe the trend for flu during the whole time period?

   _____

   _____

5. How would you describe the trend for hypothermia during the whole time period?

   _____

The table shows the graphics that you learned about in
Unit Three.

Graphics	What They Are Usually Used For
Diagram or Schematic	• It shows what something looks like and how it works.
Map	• It shows how things are arranged in a place such as a floor of a building.
Flowchart	• It shows the order of steps in a procedure.
Table	• It shows lists of details about a subject.
Line graph	• It shows how one amount changes over a period of time.

UNIT CHECK

Answer the questions.

1. Suppose you're reading a bus schedule. It lists the different times that a bus reaches different bus stops. What kind of graphic is the schedule? _____

2. A graphic shows how a computer is connected to a printer. What kind of graphic is it? _____

3. Suppose you read a graphic that shows the floor plan of a library. What kind of graphic is it?

_____

4. A shoe company uses a graphic to show how much money it made in every month of 1990. What kind of graphic is used?

_____

TALK ABOUT IT

1. Newspapers use many kinds of graphic aids. Find one graphic whose subject interests you and share it with your class.

2. Bring in real-life examples of tables, such as bus schedules, take-out menus, and tax tables. Discuss what details a table describes and whether the table is easy to understand. Come up with ways to improve the table.

UNIT FOUR

Reading to Follow Instructions

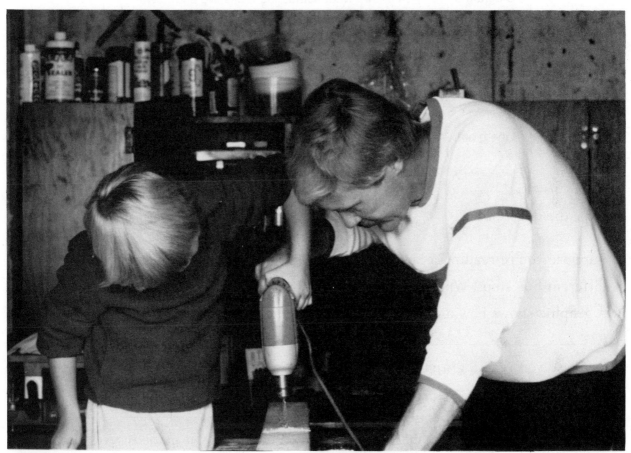
Many do-it-yourself projects come with written directions that must be read and followed.

Have you ever bought something that you had to build yourself? How well did you follow the instructions that came with the kit?

Some people read instructions and say "They don't make sense. Let's just try to figure it out." And they do figure it out eventually. But it may take them two or three times longer than if they had followed the instructions.

Trying to figure out something on your own is fine sometimes, especially when directions are poorly written and don't make sense. When you're on a job, you'll be given many written instructions. You will be expected to read and follow them exactly the way they are written.

Unit Four is all about reading and following instructions. You'll learn ways to handle short instructions as well as long, difficult ones. You can use what you learn in the workplace and anywhere else you need to follow written instructions.

LESSON 11

Reading and Following Instructions

Journal entry for January

I never knew how dependent I was on my mother and ex-wife for so many things. Today I used a washing machine for the first time. It was a disaster. I put too much soap in the machine and flooded the Laundromat floor with suds.

The owner, an old Chinese woman, scolded me at first. I guess I looked so sad that she stopped. "No more wife," she said. "No more girlfriend."

I felt myself turning red. I said, "I'm sorry. This is my first time washing for myself."

She studied me for a moment, as if she couldn't believe that a grown man had never washed a load of clothes before. Then she picked up my soap box and placed it in my hand.

"OK, I'll teach you. You read this," she said. She pointed to the directions on the back of the box. And then she walked away.

I needed only half a cup of soap. I had put in almost half the box. I should have read the directions—this from a guy who gives instructions to people at work.

The owner came back. She held a measuring cup in one hand and a mop in the other. She placed the measuring cup in my hand.

She said, "Use this to measure out your detergent."

"Thank you," I said. "Please, I'll mop the floor."

She shook her head. She smiled. "I'll mop now, but you'll mop if it happens again."

I followed her instructions and measured out the detergent. When the rinse cycle started, I put in the fabric softener. After the clothes were dried and folded, I stood back and admired the display on the folding table. Washing clothes wasn't so bad after all.

Talk about It

- What problem did the writer have at the Laundromat?

- How could he have prevented the suds from overflowing?

- Did you ever fail to follow directions?

- Did anything funny happen?

- Did anything disastrous happen?

Reading Instructions

We read **instructions** everywhere—in our homes and on our jobs, on buses and streets, in theaters and almost anyplace you can name. We find instructions on almost everything: clothing, machines, food packages, order forms, bills, and so on.

Instructions is another word for *directions*. Some instructions describe how to use something. For example, you might read instructions to operate an electric can opener.

Some instructions describe how to do something. For example, you might read instructions for getting a refund on something you bought.

Instructions give specific steps for what to do. By following them, you should get the right result.

Think of a time when you didn't follow instructions. Describe what happened.

▼ Work Out

Where would you find each set of instructions? Choose your answer from the words below.

a can of soup   **a credit card bill**
a contest form   **a bottle of medicine**

1. Take two teaspoons every four hours. Do not use for more than five days.

_____

2. Payment is due by the tenth of the month. Send only check or money order.

_____

3. Warm contents over medium heat or heat in microwave for forty-five seconds.

_____

4. Write the words Good Health on a three-inch by five-inch piece of paper.

_____

A Strategy for Following Instructions

It often helps to have a **strategy**—a plan of action—to help you understand instructions. On the following pages you'll learn about a five-step strategy that can help you.

Step 1: Define your goal.
A goal is something that you want to achieve. Before you read a set of instructions, ask yourself: What is it that I want to get done?

> To order your free jacket pattern complete the order form: Print your name and address and circle the correct size. Send the form and $1.00 (for postage and order processing) to:
>
> *The Pattern Company*
> *631 Third Street*
> *Hanford, CA 95523*
>
> Only one free pattern. You may buy additional patterns for $5.86.

▶ Read the set of instructions above. What would be your goal?

_____

Answer: The goal is to get a free jacket pattern.

Step 2: Identify the steps you need to follow.
As you read, note the steps in some way. Underline them or write them out. Then read over the steps you noted. Ask yourself: Can I complete my task by following these steps?

▶ You would complete two steps to get the free pattern. What are they?

_____

Answer: The two steps are (1) fill out the order form and (2) send it along with $1 to the address provided.

Step 3: Name the items you need to finish the task.
The instructions won't always state all the items that you'll need. But from your own experience you can tell what you may need. For example, you'll need an envelope and stamp to get the free jacket pattern.

Step 4: Clear up any details that you don't understand.
Make sure that you understand all of the instructions
before you actually do the task. If you're unsure of
anything, find someone who can help you out.

Step 5: Evaluate the way you did the task.
You do this step after doing the task. Examine the way
you did the task. Did you achieve your goal?

▼ Work Out

Read the instructions. Answer the questions based on four of the five steps:

1. Define your goal.

2. Identify the steps you need to follow.

3. Name the items you need to finish the task.

4. Clear up any details that you don't understand.

To get a library card, you must complete an application
form. You must show proof of your address. You can show
a driver's license, a passport, a personal check, a bill, or a
letter that shows your address. You must pay a $2 fee.

1. Suppose you follow the instructions. What is your goal?

   _____

2. What three things must you do to accomplish your
   goal?

   _____

   _____

   _____

3. What will you need to accomplish your goal?

   _____

   _____

   _____

4. Are there any details that you would like to understand
   better? What are they?

   _____

   _____

Applying the Strategy to the Job

You'll read many written instructions in the workplace. This strategy can help you read any kind of instructions. Let's review the first four steps of the strategy.

Step 1: Define your goal.
Below is a job order that an office worker might get.

JOB ORDER

Date: March 11, 199___

    Please make twenty-five copies of the report. Copy the pages back to back. Put each copy into a binder and make a label for each binder.

    Distribute the binders to the employees on the attached list. If you cannot complete this job by Friday, please see me.

Mel Obeso *MO*

▶ Write the worker's goal on this line.

_____

The goal is, *"Pass out twenty-five copies of a report by Friday."*

Step 2: Identify the steps you need to follow.
Note the steps in some way such as underlining them. If steps must follow a certain order, number them. Find the steps in the memo and underline them.

You should have found four steps: (1) Make twenty-five copies of a report. (2) Put the copies in binders. (3) Make labels for the binders. (4) Pass out the binders.

Step 3: Name the items you need to finish the task.
▶ What does the worker need to complete the job order?

_____

The worker would need at least these things: the report, list of employees, binders, labels, copier, and paper.

Step 4: Clear up any details that you don't understand.
▶ Can you see yourself carrying out every detail of the instructions? If anything is unclear, ask about it. Who do you think would be the best person to ask?

_____

Here's the last step of the strategy for reading instructions.

Step 5: Evaluate the way you did the task.
You do this step after doing the task. Examine the way you did the task. Did you achieve your goal?

If you didn't, go over your steps. See where you may have forgotten to do something or may have done something the wrong way. Make a note to yourself that describes what you must fix or what you should do the next time you perform the task.

▼ Work Out

If you were an office worker, you might get instructions like these:

> Irma, please type this letter on letterhead. Have Mr. Burton sign the letter. Make three copies—one for Mr. Burton, me, and the files. Send the letter out today.
>
> Ms. Williams

1. What is the employee's goal?

_____

2. What four steps must the employee take to achieve her goal?

_____

_____

_____

3. What does she need to complete the task?

_____

4. Suppose you're the worker. What details would you need to have made clearer?

_____

_____

_____

▼ Work Out

Here's a memo that workers at a community meals program might receive. Read the memo. Then read what two workers did. Evaluate how the workers followed the instructions.

To: All food service workers
From: Community Meals Director
Date: August 15, 199___
RE: Procedures for Serving Food

    Wear apron, plastic gloves, and cap or hair net when serving food. Bare hands should never touch food. Serve meal on a plate. Serve side dishes such as salad, soup, and dessert in separate dishes or bowls. Never take food to empty seats. Take food and a glass of milk to participants as they are seated at a table.

1. Frank put on a cap before serving the food. He placed rice, beef, cooked vegetables, green salad, and dessert on a plate. He took the plate to Mr. Kim, a participant. Did Frank follow instructions correctly? Why or why not?

_____

_____

_____

2. Jenna put on a hair net, gloves, and apron. She served all the food in separate dishes and plates. She placed it on the tables before the participants arrived. She escorted participants to seats as they came in. Did Jenna follow instructions correctly? Why or why not?

_____

_____

_____

ON THE JOB

Suppose you work in the shipping department of a book company. Your supervisor gets this job order. He assigns the job to you. Read the job order. Then answer the questions.

NEW BOOKS, INC.
JOB ORDER

DATE: August 28, 199___

Mark, the order for Willows School must get to the school by next Thursday, September 6th.

Part of the order is for one hundred of the new health workbooks. They are expected to come in by September 4th. Wait for the books and send the complete order to the school by special delivery.

Larry

---

1. What is your goal?

_____

2. Check each step that you would need to do.

_____ Fill the order.

_____ Tell the school that the order is on the way.

_____ Send the complete order by special delivery.

_____ Send part of the order now; then send the rest of the order when the books come in.

_____ Send the order by September 4th.

3. What materials would you need to do the task?

☐ the order form

☐ 100 new health books

☐ books that are listed on the order form

4. Is there any part of the job order that you don't understand? Write the questions that you'd ask your supervisor to clear up.

_____

_____

5. In your own words, write a summary of the task that you must do.

_____

_____

Reading and Following Procedures

Bo drove into a self-service gas station. He didn't mind pumping his own gas for the price—$1.25 a gallon. He looked at the long line in front of the cashier window. He would be late for his meeting. But he needed gas to get there. He got in line. When it was his turn, Bo said, "Eight dollars on pump three, please."

"Pump your gas first," the cashier said.

"I'm late for a meeting. Let me pay now. I stood in line for a long time."

"Sorry," the cashier said. "This is the way we sell gas. You should have read the directions."

Bo walked back to the pump. Angry, he opened his car's gas tank. He grabbed the hose from the pump and flipped a lever up. He pushed the hose nozzle into the tank and squeezed the trigger. Nothing happened.

Bo yelled at the cashier, "Start the pump!"

The cashier's voice boomed over the loudspeaker. "Follow the directions."

Bo stared at a sign. But he was too angry to understand the words.

A man at the next pump said, "Press the *Cash* or *Credit* button to start it."

Self-service stations have set procedures for pumping gas.

"Buttons?" Bo looked around for them.

The man came over. "Are you paying with cash or a credit card?"

"Cash," Bo said. "Eight dollars."

The man pressed a *Cash* button next to the money meter on Bo's pump.

"Go ahead," he said. "Watch the meter. Stop when it shows eight dollars."

"Thanks," Bo said, turning red. "Why can't all gas stations use the same procedure?"

Talk about It

• What is Bo's problem?

• How could Bo have prevented the problem in the first place?

• What do you think Bo means by his question "Why can't all gas stations use the same procedure?"

The Goal

A **procedure** is a way to do or make something. You follow procedures, for example, when you follow a recipe or instructions to build shelves.

Every business and company has procedures that people must follow. Some procedures are for customers, such as buying gasoline. Other procedures are for workers, such as completing time sheets.

Procedures are not difficult to follow. First, state the purpose for a procedure. Ask yourself: *What is my goal?* Whenever you become confused when you read or follow the procedure, remind yourself of your goal.

procedure: a particular way of doing something, usually involving ordered steps

---

▼ Work Out

Read the procedures. What is the goal for each one? Check the best answer.

1. For a cash refund, take the item and receipt to the second floor. Fill out a form; then go to the refund window. If the clerk approves, you will get a copy of the form. A check will be mailed to you within six weeks.

   _____ **a.** filling out a form for a refund

   _____ **b.** getting your money back for something that you bought

   _____ **c.** finding the refund window

2. Every Friday, tell your supervisor the hours and days that you can work the following week. The supervisor must approve your schedule. If you can or wish to work longer hours, also inform your supervisor. If approved, your supervisor will post the new work schedule by Saturday. If not approved, the supervisor must assign a schedule to you based on the company's needs.

   _____ **a.** disputing a work schedule

   _____ **b.** changing a work schedule

   _____ **c.** getting a new work schedule

The Right Order

To do a task correctly, you must follow steps in the right order. When the steps are not numbered, you have to figure out where one step ends and another begins.

Here's a tip that may help you: find the main action that occurs in a step. Sentences that start with verbs are often clues that a new step is beginning.

▶ Read this procedure for making hamburgers in the microwave oven. Find the five steps to it. Underline the first word that starts a new step.

Mix ground beef with spices. Spices are of your choice. Shape meat into six patties. Place patties in a baking dish and cover it with waxed paper. Microwave for 6 to 8 minutes. The power setting should be *High*.

The five steps in the sequence start with these words: *Mix, Shape, Place, Cover, Microwave*. All five words are verbs. They each describe the main action in a step.

▼ Work Out

Here's a procedure involving a public bus system. It has five steps. Read the procedure; write *1, 2, 3, 4,* or *5* to show the correct sequence.

_____ Take a transfer from the driver.

_____ Pull the cord to signal that you'll get off at the next stop.

_____ Put your fare in the collection box.

_____ Get off the bus at the back door.

_____ Find a seat for yourself.

Reading Procedures on a Flowchart

Sometimes procedures are written on flowcharts like the ones you studied in Lesson 8.

Symbols stand for the steps. Arrows tell you in which order to follow the steps.

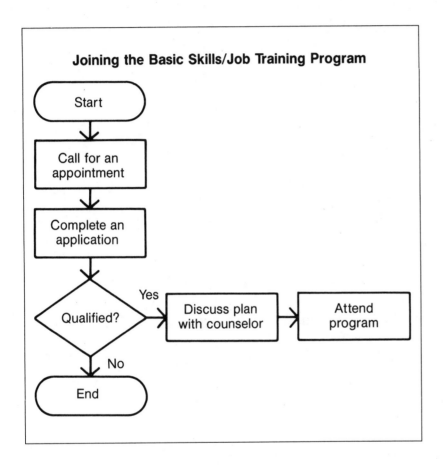

Joining the Basic Skills/Job Training Program

Start → Call for an appointment → Complete an application → Qualified?

Qualified? — Yes → Discuss plan with counselor → Attend program

Qualified? — No → End

1. Call office for an appointment. Bring a picture ID, social security card, and last year's 1040 tax form.

2. Complete an application.

3. It takes two weeks to find out if you are qualified to be part of the program. We will mail you a letter with the results.

4. First-time participants will meet with a counselor. Together you will make a program plan that fits your needs.

5. Attend your classes, workshops, and/or job training. Only you can make it a success.

You probably noticed that the flowchart is a little different from the ones you studied in Lesson 8. Here, as in the earlier ones, the flowchart is used to sum up a procedure. The main point of each step is written in a symbol. However, clearer and more exact instructions are shown below the symbols.

Look at the written instructions. Each step is labeled with a number. While reading the instructions, label the symbols on the chart. Write *1, 2, 3,* and *4.* (Remember to follow the arrows for the sequence.)

▶ Now read the whole procedure—the symbols and written instructions. Then answer the questions.

1. Imagine that you'll follow the procedure. What is your goal?

   _____

2. In your own words, write each step. Make sure that the steps are in the correct sequence.

   a. _____

   b. _____

   c. _____

   d. _____

   e. _____

You should have written that your goal is to: *join the basic skills/job training program.* The steps include: *a. calling for an appointment, b. completing an application, c. finding out if you qualify, d. being interviewed by a counselor,* and *e. taking part in the program.*

Making a Flowchart

1. Begin by drawing the symbol *start*.

2. Draw the symbol of the first step, and write the step inside.

3. Continue writing steps inside symbols. Draw arrows between symbols to show the sequence of steps.

4. When you have finished the sequence, make your last symbol (end) .

5. Give your flowchart a title.

▼ Work Out

Here's a procedure an office worker has to follow. It describes how to turn on a computer. Read the procedure; then make a flowchart for it. The symbols are drawn for you, and some of the steps are filled in.

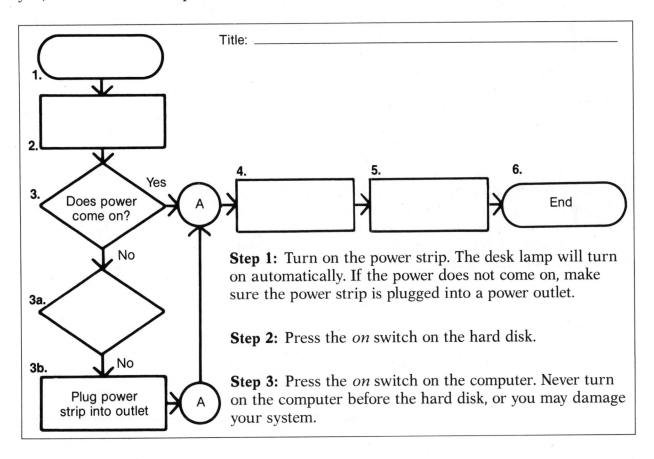

Step 1: Turn on the power strip. The desk lamp will turn on automatically. If the power does not come on, make sure the power strip is plugged into a power outlet.

Step 2: Press the *on* switch on the hard disk.

Step 3: Press the *on* switch on the computer. Never turn on the computer before the hard disk, or you may damage your system.

ON THE JOB

In the space below, make a flowchart with written instructions for a four-step procedure. Choose a task that you do at home, school, or work. Here's what to do:

1. Write a title for the procedure. It should answer this question: *What goal do you want to accomplish?*

2. Write the steps for the procedure. Number each step.

3. Make the flowchart. Write the main point of each step in a symbol. Draw arrows to show the correct sequence of the procedure. Refer to the flowchart on pages 58, 60, and 88 if you need to.

Instructions explain how to make or do something. Instructions that give exact steps in a certain order to accomplish a goal are called **procedures**.

The following strategy can help you read and follow any set of instructions or procedures.

Step 1: Define your goal.

Step 2: Identify the steps you need to follow.

Step 3: Name the items you need to finish the task.

Step 4: Clear up any details that you don't understand.

Step 5: Evaluate the way you did the task.

UNIT CHECK

Imagine that you are a teacher's aide. You get these instructions from a teacher. Read them. Then answer the questions.

Please correct Group A math tests for periods 1, 3, and 6. Enter test results in my grade book. I would like to get the tests back to the students tomorrow.

1. What would be your goal?

_____

2. What two steps would you do in the task?

_____

_____

3. What things will you need to do the task?

_____

_____

4. Write the instructions over in your own words.

_____

_____

_____

_____

TALK ABOUT IT

1. Find an example of poorly written instructions or procedures. Discuss what's wrong with them and how you would improve them. Then rewrite the instructions or procedures.

2. Choose a card game that you like. Write instructions for it and teach your class how to play it.

3. Choose a room in your home such as the kitchen or bathroom. Find things that have instructions on them. Copy the instructions or cut them out. Bring them to class. Discuss how well they are written and ways in which you might improve them.

Finding Useful Information

Imagine that you need the phone number for the *Salinas Insurance Company*. You have a phone directory. How would you find the number? Would you read every page in the directory until you come to the *S* section? Would you then read every entry until you come to *Salinas Insurance Company*?

No. Most likely you would turn to the *S* section right away. Maybe you'd use the guide words at the top of the pages to help you find the right page. Then you'd glance through the page until you come to *Salinas Insurance Company*.

Finding information in a phone directory is an example of scanning. Scanning and skimming are two reading skills that can help you find information quickly. When you scan or skim, you're not reading every word in the text.

These two reading skills can be very useful in the workplace. For example, let's say you need to review a certain step in a procedure. You should not have to read the whole procedure just to find that step, and you may not need to read that entire step either.

You'll learn how to get information from reading materials in this unit. Lesson 13 will show you how to scan parts

You scan when looking for names and numbers in a phone book.

of a manual to see if it has the specific information you need. Lesson 14 will show you how to skim text for general information.

Using a Reference Tool

Lourdes watched her husband, Roy, flip through the catalog for the third time, looking for the camping gear.

The thin pages were sticking to his sweaty palms.

"Do you remember where you saw the camping gear?" he asked his wife. "I thought it was in the back with the basketballs."

"Why not try looking in the index?" asked Lourdes.

"You've got a point," Roy said. "I guess they're included in catalogs for a good reason." He looked under the heading *Gear* but with no luck.

"It's not under *gear*—there's no listing for camping *gear. . .*"

"Honey, let me take a look," Lourdes said.

"I'll find it," Roy replied. "I'll try *equipment* and see if it's there."

He flipped to the heading that read *equipment.* Under the heading appeared a long list of entries. He ran his finger down the listing until he found the word *camping.*

Finally he turned to the page number on which it was listed.

A table of contents is very helpful when looking for items in a catalog.

Talk about It

- What was Roy looking for in the catalog?

- What did Roy do that was easier than flipping through the pages?

- How do you look for specific information in a catalog, manual, or other book?

Using the Table of Contents

Many books discuss a number of different **topics**. You can learn what the topics are in the book's **table of contents**.

At the right is an example of a table of contents. It shows the topics that might be discussed in a community directory.

A table of contents also lists the other parts of a book, such as the introduction and the index.

Sometimes a topic is broken down into more specific topics. A table of contents may list the specific topics as well.

When you want to be sure a book has the information that you need, you can **scan** the table of contents.

▼ Work Out

Refer to the table of contents above to do this exercise. The first question has been done for you.

1. You need to make a will. Where could you find a lawyer that charges low fees?

   Legal service information

2. You want to enroll in computer classes. Where could you look for information?

   _____

3. You need to go to the public library. Where could you find information on its hours?

   _____

4. You're taking someone special out to dinner. That person would like to eat Chinese or Italian food. Where could you look to find a restaurant near your home?

   _____

5. You're going to a shopping center. You'll take a bus. Where should you look to find out what bus you should take?

   _____

Using the Index

When you must find specific information in a book, the *index* can help you find it more easily.

Remember, the index is in the back of a book. It's a list of topics, as well as names and titles, that are discussed in the book. The index lists things in alphabetical order.

At the right is an example of what an index looks like. It's part of an index that you might find in a secretary's handbook. The **entries** are for the letter *W*.

Look at the index. Notice that some topics are broken into more specific topics. The page number (or numbers) where a topic is discussed is listed. Some topics are covered on one page only. Others are discussed on several pages.

▶ Scan the index for the entry *Worker's compensation.* On what page or pages is the topic discussed? _____

▶ Scan for the entry *Word processing function keys.* On what page or pages is it discussed? _____

Did you answer *pages 333–334* and *page 68?*

Sometimes a topic is **cross-referenced**. That means you can get more information about the topic under another topic. When a topic is cross-referenced, you'll read the word *See* next to it.

Some indexes also show if an illustration—picture or graphic—is on a page. Then the word *illustration* or *illus.* (its abbreviation) will appear next to the topic.

▶ Which topic includes an illustration? On what page is the illustration found?

Did you answer *WATS, page 193?*

W

Wages. *See* Salary.
  in union contracts, 310
WATS (Wide-Area
Telecommunications Service)
  lines, 187–191, 193 (*illus.*), 195
Western Union, 215–232
Word processing, 42–71.
    See Dictation.
    career, 12
    electronic typewriters,
      40–41
    health, 22–23
    security systems, 110
Word processing equipment,
  care of, 128–130
Word processing function
  keys, 68
Word processing systems,
  50–55
Worker's compensation,
  333–334

Where Do You Find It?

Sometimes you'll look up a topic that's not listed in an index. But you know the topic is discussed in the book. What do you do? When that happens, you must think of other ways that the topic may be listed.

Let's say you need to know the best way to write a telegram. You scan for *Telegram*. No listing. You scan for *Writing telegrams*. No listing there either. Then you scan for *Western Union*—that's the place that sends telegrams. There's a listing. You turn to the right page and scan the text for the information that you need.

▼ Work Out

Use the index on page 96 to answer these questions.

1. A small office needs a word processing system. Burt, the office manager, must decide which is better: an electronic typewriter or a computer system. What topic in the index might he choose to look up?

   _____

2. Ms. Domingo owns a small clothing store. She plans to hire more part-time, experienced clerks. She needs an idea of how much to pay them. What topic might she look up?

   _____

3. Bill is a word processor. He can't remember what function key to press on his machine to center a line of copy. The index has no listing for *Copy* or *Centering copy*. What entry might have the information that he needs?

   _____

4. Terry's job is to take care of the word processing equipment. He's been told that the room it's in must stay at a certain temperature. The index doesn't have an entry for *Temperature for word processing equipment*. What entry might have the information?

   _____

Parts of a Manual

The crossword puzzle below provides a review of the kinds of information that can be found at the front and back parts of a manual or book.

▼ Work Out

Complete the crossword puzzle. Use the words to the right to help you.

appendix   illustrations
contents   index
copyright   introduction
glossary   title

Across

3. To find out who wrote a manual, you'd look at the _____ page.

5. You can learn what topics are in a manual from the table of _____.

6. To make sure you have the most recent copy of a manual, you'd read the year it was published on the _____ page.

7. You can find the page that a subject is discussed in this part of a manual.

8. This part usually holds extra information, such as graphics and documents that supplement the text in a manual.

Down

1. When you want to know if a manual has a certain graphic, you can look through the list of _____.

2. You can learn the meaning of a technical term in this part of a manual.

4. By reading this part, you get an idea of the purpose and the subject matter of the manual.

ON THE JOB

Imagine that you're hired by a large company. On your first day of work you get an employee's handbook. Its table of contents is shown below. Use it to help you answer the questions.

Table of Contents

1. You want to know who is in charge of the different departments in the company. What section would give that information?

_____

2. You want to know how long a new worker is on probation, is on a trial period. Where would it be covered in the employee's handbook?

_____

3. Suppose you'll work with dangerous chemicals. The handbook states what kind of clothes you must wear. Which section would talk about that information?

_____

4. You want to know the company's policy on personal telephone calls. You don't find any listing in the index. From the table of contents, where might you look?

_____

Reading Quickly for Information

Marva stared at the mess on the coffee table. There was no way around it. Today she would have to read the mail. She had put it off long enough.

When Marva was in the hospital, nobody had opened her mail, and during the two weeks since she'd been home she had lacked the energy to do anything. She had asked Beryl, her oldest child, to look at the mail. But a ten-year-old could understand only so much.

Marva sat on the couch, pulling the coffee table closer to her. She went right to work. She made separate stacks for magazines, advertisements, bills, letters from people she knew, and letters from people she didn't know.

Marva skimmed through letters asking for donations. She put some letters aside. Others she threw away. She read personal letters more slowly. Opening bills with a frown, she looked only for the amount that she had to pay. The bills that were past due required immediate attention. The others would have to wait.

She flipped through the magazines, marking the pages that she would take the time to read later.

As Marva finished the last piece of mail, she heard someone rattling a key in the lock. She glanced at her watch—3:30. Soon the apartment filled with the noise and movement of three small children.

"What a mess, Mom," said Beryl. You've got mail all over the place." Beryl started to organize the envelopes and magazines that her mother had scattered across the coffee table.

Marva said, "I deserve a treat, Beryl. I finished reading all the mail."

"You forgot this," Beryl said, dumping the day's mail on her mother's lap.

Marva groaned. The three children laughed. They brought flowers from behind their backs.

"It's good to have you home, Mom," they said.

"Now this is a treat," said their mother.

Talk about It

- Which kind of mail did Marva take her time in reading?

- Why do you think she read this mail more carefully than the rest?

- Why do you think Marva was able to read through all of her mail in one sitting?

- Suppose you had a large pile of mail. What would you do to finish reading everything in one sitting?

Scanning or Skimming?

When the purpose of your reading is to find certain information, you can either **scan** or **skim** the materials.

Skimming and scanning skills help you to read words quickly. When you're skimming, your pace is faster because your goal is not to read every single word.

When you want specific facts such as a name or date, you *scan* the text. When you want a general idea about what's being discussed in the text, you *skim* it. For example, you skim a few paragraphs of a magazine article to see if it interests you enough to read it all. You scan a store ad to check the sale prices for a specific item that you know is on sale.

To help you scan and skim, keep your purpose in mind.

scan: to read quickly to find specific information

skim: to read quickly to get the main idea

▼ Work Out

Answer the questions. Decide when you would skim or scan reading material.

1. Suppose you're reading a newspaper. You see an article about jobs. You wonder if it might have useful information. How would you read the article? Why?

   _____

   _____

   _____

2. Let's say you're going to prepare a chicken recipe. You need to know how much chicken to buy. How would you read? Why?

   _____

   _____

3. **a.** Suppose you want a part-time job as an office worker. How would you read the "help wanted" ads? Why?

   _____

   _____

   b. To see if the job is something that you'd like to do, would you skim or scan an ad? Why?

   _____

   _____

Skimming through Text

Remember, you use skimming skills when you want a general idea of what is said in articles, letters, notices, memos, reports, and other reading materials. Here are some tips that you may find helpful:

- **Know what the main point is for the complete text.** The main point is usually given in the first one to three paragraphs. Read them carefully. Also read the last few paragraphs of the text. They usually sum up the main point, too.

- **Skim through the rest of the text.** Look for the main point in each paragraph. One sentence in each paragraph often will sum it up. Usually that's the first or last sentence of a paragraph.

- **Read headings.** They are clues to what's covered in a section. Also read explanations of sentences that are underlined or printed in *italics* or **bold** type. They often represent important facts that the author wants the readers to remember. Finally, examine all graphics that appear with the text.

- **Find a skimming speed that's comfortable and fast** enough so that you're not actually reading every word, yet slow enough so that your mind is making sense of the words. Some people find it easier to concentrate by sweeping their finger or an index card across the page to lead their eyes.

  These tips can help you skim through graphics as well. You don't want to skip over the information that a graphic has, because graphics sum up in pictures what has been said in words.

▶ Skim the following, and write the main idea of the passage.

> Many companies are now offering flextime to their employees. Flextime allows workers who are parents or who are attending school to arrange their 8-hour workday around their other activities. Instead of working a 9–5 hour day, workers may work 7 to 3 or 10 to 6. In other cases, two people may share a 40-hour a week job.

You may have written for the main idea: *Many companies are offering flextime to help their workers* who may be parents or students.

▼ Work Out

Practice skimming short pieces of text or graphics. Answer the questions after you skim each piece.

Read the job description below and answer the questions that follow.

> Every ten years the United States takes a census, a count of the population. One of the key jobs at census time is enumerating, or counting, the population. Enumerators meet persons face-to-face. They go into neighborhoods, group living quarters such as nursing homes and prisons, and other places where people may live.
>
> Some enumerators develop address lists by going out and checking addresses. Some enumerators distribute and collect census forms from households. If for some reason members of a household do not complete a form, other enumerators must interview them. Some questions that are asked are very personal.
>
> Enumerators do a lot of walking and climbing on their jobs. Often they must work evenings and weekends to talk with persons who are home only during those times.

1. What is the main point of the text? Check the best answer.

_____ **a.** The census is taken every ten years.

_____ **b.** An enumerator gathers information for the census through a variety of methods.

_____ **c.** Census workers must like to meet people.

2. What work does one kind of enumerator do?

_____

3. Where do enumerators work?

_____

4. Would you apply for a job as an enumerator? Why or why not?

_____

_____

ON THE JOB

Imagine that you're in a workplace. You get the memo below. Find the main point of the memo by skimming it. Then answer the questions. Find the answers by scanning.

To: All employees
From: The President
Date: September 23, 199___
Subject: New procedures for visitors

As of October 1, 199___, new procedures for visitors to the company will begin.

All visitors—personal and business—must wait in the lobby. An employee must escort them to and from the lobby. A visitor must *always* leave the building by way of the lobby.

The receptionist will call the proper employee about a visitor. The receptionist will record the time that a visitor enters and leaves the building.

If an employee expects to give a company tour, it must first be cleared with his or her supervisor. At least two days' notice must be given. That way department managers will be aware of possible visitors in their area.

Should you have any questions about the new procedures, please refer them to your supervisor or the Personnel Department.

---

1. The main point of this memo is: _____

_____

2. When must you start following the new procedures?

_____

3. Suppose a service person comes to repair the copier in the accounting department. What must the receptionist do?

_____

4. Let's say your supervisor had a visitor. He asks you to show the visitor out of the building. Where do you take her?

_____

5. A relative visits you at work. She would like to tour the company. Could you take her around? Why or why not?

_____

UNIT FIVE SUMMARY

To help you find certain information in a book, use either the table of contents or the index. From the table of contents you can get a good idea of the subject matter in a book—what the major topics are and the sequence in which they will be discussed. From the index you can learn where specific topics, persons, or titles are discussed.

By skimming books, letters, and other reading material, you can get a general idea of their contents. When you need to find specific information in reading material, use your scanning skills.

UNIT CHECK

Answer the questions.

1. How can you use the table of contents to help you skim through an employee handbook?

_____

_____

_____

2. How can you use an index to help you scan for information in a procedures manual?

_____

_____

_____

TALK ABOUT IT

1. What's the difference between reading for pleasure and reading to do something?

2. Do you think a fast reading rate helps a person understand better?

3. How might a lack of skimming and scanning skills cause a person difficulty on a job?

Problem Solving in Life

Employers reward workers who are good at solving work-related problems.

Imagine yourself getting an award from an employer. In a speech he says, "You are a hard-working person. You do your best at all times. You handle problems in a sensible and calm manner. A hard worker and a problem solver, that's you."

Employers like workers who can handle problems. Problems are always popping up in the workplace. Machines stop working. Electrical power goes out. More work comes in before the workers finish the last job. A work accident hurts several workers in ten seconds. The kinds of problems that can arise are many.

Most of us probably get nervous when we think about dealing with problems. We don't know where to begin. That's when having a strategy or plan can be useful. It can help us focus our thoughts and energy on the problem.

In this unit you'll learn about a strategy for solving problems. The strategy can be helpful to you in any workplace.

You'll see how the strategy can help you:

- handle situations in which a procedure isn't working successfully

- plan ahead to deal with an ongoing problem

- cope with emergencies

A Problem-Solving Strategy

Tania Williams stared at the pile of bills. Every month she and her four kids just managed to get by. She always paid the rent, gas and electricity, and phone bills. If she missed a payment to a store or credit card, well, she missed it. It couldn't be helped.

"What's the matter, Mom?" said Ernie, her sixteen-year-old son.

"I'm tired of robbing Peter to pay Paul."

Ernie noticed the bills. "I'll be getting my paycheck on Friday, Mom."

"Thanks, Ernie," Tania said.

She was grateful that Ernie gave her most of his pay when she hadn't asked him to. She picked up a bill and sighed.

"These bills never seem to go down."

"You know, Mom, I'm in an economics class. Our teacher has us pretending we're owners of a small business. Last week he gave us a problem kind of like yours. We had to figure how to change things so that we could pay a big debt and start making money.

"What if we pretend that our family is a business, Mom? You're the owner. I'm your consultant. I'll help you solve your money problem."

"All right, Mr. Consultant. Where do we start?" said Tania.

Managing the monthly bills can be a form of problem solving.

"We start by stating your problem."

"That's easy. This business has no money at the end of the month."

"I'll go get my notebook and economics book. Mom, you think about how much you'd like to have at the end of the month."

Tania smiled proudly at her son. Maybe he can come up with a plan that works, she thought. Then I could pay *all* my bills. Maybe we can start saving money.

She began to feel better.

Talk about It

- What is Tania's problem?

- What does Ernie suggest that they do to tackle Tania's problem?

- Why is making a plan to solve a problem helpful?

What's the Problem?

On the following pages you'll learn the steps to a strategy that can help you handle problems in any situation.

The steps are:

Step 1: Define the problem.
Step 2: Make up a list of fact-finding questions.
Step 3: Gather your information.
Step 4: Analyze your information.
Step 5: Examine possible solutions.
Step 6: Choose the best solution.

To use the steps, let's look at the story of Tania Williams. Several things worried Tania. She had a job, but she couldn't pay some bills each month. The bills never seemed to go down. Her family never had money at the end of the month. They couldn't save money.

Step 1: Define the Problem.
You must always have in your mind what it is you're solving. So the first step is to explain what the problem is. Be as specific as you can. To keep yourself focused, state the problem as a question. The question should suggest what it is you want to solve.

How could you define the problem? You could sum up Tania's problem as: **She was not able to pay her bills and save money.**

Step 2: Make up a list of fact-finding questions.
Now you need information about the problem. To focus your search for facts, think of questions that ask for facts as answers. For example, Tania might ask: What does each bill go toward? How much is each bill? When must I pay each bill? How do I decide which bills not to pay? How much money does our family bring home?

▶ What are two other facts that she might need to know?

_____

Two other facts that she might ask for are: *Which things do we buy with cash and which with credit cards? How much money would I like to save each month?*

Finding the Information

Once you have your list of questions, you're ready for the next step of the problem-solving strategy.

Step 3: Gather your information.
Organize the way you gather your information. First, take the time to think about the sources that may have information you need. What written materials such as bills or letters, reports or articles, or books could help you? Make a list of the materials; then go get them.

Remember the problem Tania wants to solve: *How can she pay the bills and save money each month?*

The sources checked below might give her information that she needs.

☑ bill statements
☑ articles on: saving money
☑ articles on: managing money
☑ checkbook

☑ pay stubs
☑ articles on: making a budget
☑ articles on: paying with credit cards

Remember to use the skimming and scanning skills that you learned in Unit Five. When you come to text that may have the facts you need, slow down and read carefully. As you read, you must decide whether the information fits your needs. Does it answer any or all of your questions? You may want to take notes to help you remember all the facts.

Let's say you go to the library to find information on managing money. You find two books that may be useful. If you look through the table of contents, you can decide which book to skim.

The table of contents will list the chapters or topics that each of the books contains. Based on this listing of topics, you can immediately see which of the books best meets your needs. Then you can eliminate the one that does not contain information useful to you. In this way, reviewing the table of contents of a book can save you time.

Finding a Solution

Here are the last three steps of the problem-solving strategy.

Step 4: Analyze your information.
Go over all your facts. What do they all mean? You should be able to see the causes of your problem.

Suppose, after reviewing her information, Tania sees this:

Except for rent and phone, gas, and electricity bills, the family pays for most things with credit cards. The family eats out too much. The Williamses buy too many clothes. And they buy things that they don't really need.

Step 5: Examine possible solutions.
A **solution** is the answer to your problem. Most problems can be solved in a number of ways. So think of as many possible solutions as you can. If you can, actually try out the solutions. What seems to be a solution may not always work in real life.

Step 6: Choose the best solution.
Keep in mind that a solution may take time to work. Tania, for example, decides to try this solution:

The family will pay for everything in cash or by check. The Williamses won't use any credit cards until all the bills are paid up.

▼ Work Out

Read about a problem that a secretary has. Then answer the questions.

Fran was hired as the secretary/bookkeeper/receptionist for a small office. That was five years ago. Now that the business has grown, the owners have decided to hire an additional full-time and part-time worker. Fran is happy about that. But here's her problem: **Fran has to decide what tasks each worker will be responsible for.**

1. Check the information that would help Fran solve her problem. Write other questions that may also help her.

   ☐   What are all the tasks that Fran does?

   ☐   How often are these tasks done?

   ☐   What tasks would be done best by Fran?

   ☐   What tasks would be done best by a full-time worker?

2. Check the best sources that may help Fran. If you can think of other sources, write them below.

☐ a secretary's handbook

☐ office phone log

☐ sales reports

☐ office procedures

☐ past job orders

☐ customer invoices

☐ _____

☐ _____

3. From the information Fran gathered she makes this evaluation.

Her more difficult tasks are bookkeeping, preparing the payroll, typing reports and correspondence, and doing research. Her least difficult tasks are filing, copying, answering phones and greeting visitors, opening mail, making appointments, and running errands. She does all of the tasks every day.

a. The part-time worker would work twenty hours a week. What are some tasks that the worker should do?

_____

_____

_____

b. The full-time worker would work eight hours a day. What are some tasks that the worker should do? Why?

_____

_____

_____

A Problem-Solving Strategy

Step 1: Define the problem.

Step 2: Make up a list of fact-finding questions.

Step 3: Gather your information.

Step 4: Analyze your information.

Step 5: Examine possible solutions.

Step 6: Choose the best solution.

ON THE JOB

Suppose you're a library clerk. The librarian asks you to help her solve the problem described below. Read the problem; then answer the questions.

Almost all of the library shelves are full. Every week the library gets at least twenty-five new books. The librarian figures that in three months there may be no more space. The library has no funds this year to get new shelves. The problem you must help solve is: **What can the library do to make room for new books that come in?**

1. What fact-finding questions might you ask?

_____

_____

_____

_____

_____

2. What sources might help you find the information that you need?

_____

_____

_____

3. The librarian makes this report from the information that has been gathered.

The library shelves can hold 250,000 books. About 25,000 of those books were published between 1900 and 1940. Many of those books are rarely borrowed. The library gets between 1,500 and 2,000 new books each year. Most of the books are fiction. The cost for a set of shelves and putting them in is $1,200. The library could use at least 100 new shelves.

   a. What are two possible solutions that the library might try?

_____

_____

_____

_____

   b. Which solution do you think is the best? Why?

_____

_____

Finding the Cause of a Problem

Guests are coming over. And they should not see dust and crumbs on the floor.

I get out the vacuum cleaner. Each time I use it, I can hear my mom say, *Take it to a shop and get it cleaned.*

She said that three years ago. I never took it in. It picks up the dirt. Until today, that is. Now nothing is getting picked up.

No problem, I figure. It has to be the dust balls blocking the brush attachment. I clean it out and start the machine. No go.

Maybe the bag is full. I open up the machine. Yes, it is without doubt full. Dirt has split it open in fact. I wonder when I last changed it.

I lose twenty minutes hunting for the bags and another ten figuring out how to put in a new bag. The machine's manual, which I found in my hunt, lies at my feet. I have no time to read. Guests are coming.

I start the machine. Still nothing is picked up. I could scream, but I don't. The walls in our apartment building are too thin.

I hope something is stuck inside the hose. I go get the broom, then take off the hose. I push the broom handle through the hose. There's nothing inside it.

The manual stares back at me. I pick it up and glance at its table of contents. Great. It lists a troubleshooting chart. I look it up.

It shows a number of things to check. I have no idea what some of the things are.

No time for wondering. I look for a diagram and start checking each thing, one by one. Finally I find the problem. Something called the air outlet was blocked. I fix it and turn on the switch. The machine is going.

I start to vacuum the place. I look behind me and see dust blowing out of the machine. So, I think, when things go wrong, they get worse. Another problem! The doorbell rings. The guests have come.

Talk about It

- What problem is the character having with her vacuum cleaner?

- Why wasn't the vacuum cleaner working?

- How could the character have used the problem-solving strategy that you learned in Lesson 15 to help her?

When Something Stops Working

On the last page you read about someone who had trouble with her vacuum cleaner. She did some problem solving to get it to start working the right way. She thought of things that might have caused the trouble and went about fixing them. It took her several tries before she found the problem.

To fix something that goes wrong, you need to answer three major questions:

- What is the trouble with the machine?
- What caused it to break down?
- How do you fix it?

In the story, for example, the trouble with the vacuum cleaner was that it wasn't picking up dust. A certain part—the air outlet—was blocked. To fix the problem, the operator had to clean out the air outlet. Once that was fixed, the vacuum cleaner started picking up dust again.

▼ Work Out

Think of a time when you were using a machine. The machine stopped working the way that it should. You got the machine fixed. Answer these questions about the machine.

1. What was the machine?

_____

2. What was the trouble with the machine?

_____

_____

3. What caused the machine to stop working?

_____

_____

4. How was the machine fixed?

_____

_____

Understanding How Something Works

To find out what's wrong with a machine, you must understand how it works. Here are three important things that you should know:

1. **What a machine does.** When something goes wrong, review what the machine is supposed to do. That can help you gauge how poorly it's working.

2. **How the machine does its job.** Many of us operate machines without really knowing how they work. We know that when we turn on a switch the machine goes to work.

   We turn a key, and a car starts up. We press a button, and a copier copies a letter. We flip a switch, and a vacuum cleaner starts picking up dust.

   To find out what's wrong with a machine, you must know what is happening when its switch is turned on. You can learn that information easily. Find someone who knows about the machine to explain it to you. Or read the operator's manual that comes with the machine.

3. **What its parts are.** Think of a machine as a system of parts. All parts do a certain job. And all the parts work together so that the machine can do its job.

   You can get a quick overview of a machine by studying a graphic. Most operator's manuals have diagrams and schematics—explanations of how a thing works—that show what a machine looks like. Read a diagram first to get an idea of how the machine works. Then read it again, this time focusing on the job that each part does.

   The tips on this page and throughout Lesson 16 focus on mechanical problems. But you can use all of the information to help you handle breakdowns in any kind of system. And that includes finding out what happened in a set of procedures that gave you poor or incorrect results.

On this page is a diagram of a vacuum cleaner. Study the diagram and read the explanation below that describes how a vacuum cleaner works.

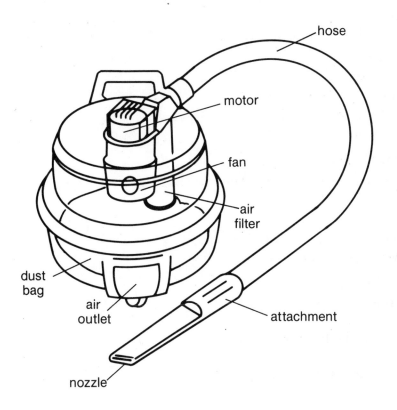

When the motor is on, it turns the fan. The fan creates a vacuum by removing the air within the machine. This causes air outside the machine to rush into the nozzle. Dust, hair, and other things get pulled into the machine by the air.

The air and dust move through the hose. The dust goes into the dust bag. Air rises through the bag and the air filter into the area where the fan is. The air leaves the machine through the air outlet.

▼ Work Out

Let's say you are planning to repair a vacuum cleaner. You study the diagram above to help you understand how the machine works. Read the diagram; then answer the questions.

1. Why can a vacuum cleaner pick up dust, hair, slips of paper, and other things?

   _____

   _____

2. Describe the path of the air as it travels through the machine.

   _____

   _____

3. What could happen if the hose, nozzle, fan, or air outlet got blocked?

   _____

   _____

4. Would the vacuum cleaner work well if the dust bag became full or the air filter became dirty? Why?

   _____

   _____

Locating the Problem

When you're locating the cause of a mechanical problem, the first thing to do is to get the right reading material. For a machine, that would be its operator's manual. Why do you think that is true?

There are three things to consider when locating the cause of a problem in a machine (or other kind of system).

- Is it operating the way it should?

  Some people think right away that a machine has stopped working because a part is damaged. But machines may stop working for very simple reasons. The power cord may be pulled out. Or the machine is out of gas, water, or oil. Sometimes a machine just needs to be cleaned.

  Let's say you want to check the oil level in a car. You might look up the topic *checking the oil level* in the index of the car's manual.

- Are you using it the right way?

  Sometimes the real problem is not the machine. It's the way a person uses the machine. You may be using the machine for jobs other than what it was built to do. You may not be following the correct procedures for operating it. Skim through the machine's manual to be sure you're using the machine the right way.

- Is something wrong with its parts?

  The machine's manual comes in handy when you must learn which part may be the problem. Most manuals have a graphic that defines possible problems, their causes, and the solutions. The graphic is called a *troubleshooting chart.*

  When you don't know what a part is, you can easily learn that too from the manual. You can look for a diagram or schematic.

Once you've found the problem, you're ready to tackle the next question: *How do you fix it?* If the suggestion that the trouble-shooting chart offers is simple, you can probably do it yourself.

Problem
Vacuum cleaner is picking up dust poorly.

Possible Cause

a. Nozzle is blocked.

b. Hose is clogged.

c. Hose is leaking.

d. Dust bag is full.

e. Attachment is cracked.

f. Fan is blocked.

g. Air outlet is blocked.

h. Motor is damaged.

Problem
Vacuum cleaner
is picking up dust poorly.

Possible Cause	Clean it out	Repair it	Replace it
a. Nozzle is blocked.	●		
b. Hose is clogged.	●		
c. Hose is leaking.		●	●
d. Dust bag is full.			●
e. Attachment is cracked.			●
f. Fan is blocked.	●		
g. Air outlet is blocked.	●		
h. Motor is damaged.		●	●

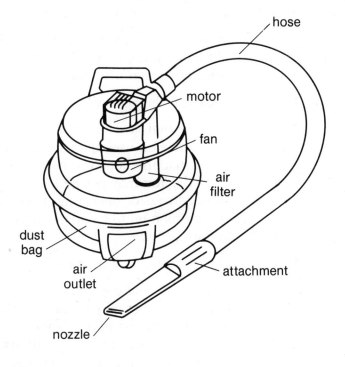

▼ Work Out

Use the graphics and the troubleshooting chart to answer these questions.

1. Mr. Torres takes his vacuum cleaner to the shop. Its motor will not run. When the service technician turns it on, the motor runs. She cleans the machine. She tries the machine, and it cleans the way it should. What could be one possible reason that the machine didn't work for Mr. Torres?

   _____

   _____

2. A service technician returns a vacuum cleaner to its owner. She tells the owner that four parts can get blocked or clogged easily. She shows the owner where the parts are and what to do if one of them gets jammed. Read the trouble-shooting chart to learn what the parts are. What are the parts?

   _____

   _____

3. Suppose you use the trouble-shooting chart to help you find the trouble with your vacuum cleaner. Read each possible cause shown on page 118. Find the corresponding part on the diagram on page 118. Write the letter (a, b, c, and so on) on the part identified as the cause.

4. What would you do if:

   a. the hose is leaking?

   _____

   b. the air outlet is blocked?

   _____

   c. the dust bag is full?

   _____

   d. the motor is damaged?

   _____

Using the Problem-Solving Strategy

You can use the problem-solving strategy that you learned in Lesson 15 to help you find the cause of trouble in a machine. Here's how.

Step 1: Define the problem.
In this case you define the trouble. Describe exactly how you were operating the machine. Also describe what the machine was doing.

Step 2: Make up a list of fact-finding questions.
Ask questions that help you understand how a machine works: What is the machine's job? How does the machine do its job? What are its parts? How does each part work?

Ask questions about the way it works: Is it installed correctly? Does it need gas, water, or oil? Does it need cleaning? Ask questions about the way you use the machine: Are you using it for what it was built for? Did you follow the correct procedures?

Step 3: Gather your information.
The machine's manual is probably the most important written source to have on hand. Remember, troubleshooting charts and diagrams can help you decide what to check in the machine.

Step 4: Analyze your information.
The main question here is: Why did the machine stop working the way it should? Most likely in this step you'll find the cause of the trouble.

Step 5: Examine possible solutions.
You're really looking at two kinds of solutions. One solution is for fixing the problem. You must decide whether you can fix it or have someone else fix it. If you can fix it, what must you do with the part, such as change it, clean it, or remove something from it?

You also need to find a long-term solution for this: Can something be done so that the trouble doesn't happen again? Examples: Clean the parts more often. Work more slowly. Use the machine for certain jobs only. This is called *preventative maintenance*, preventing a problem from starting.

Step 6: Choose the best solution.
Remember, choose the best solution for fixing the machine *and* for making sure the trouble doesn't happen again.

ON THE JOB

Choose a machine that you use often. It may be a machine in the workplace or one that you own. Get the instruction manual for the machine. Then answer the questions below.

1. What is the machine?
   _____

2. What is the name of the manual?
   _____

3. What job or jobs do you use the machine for?
   _____
   _____

4. How does the machine work?
   _____
   _____
   _____

5. On what pages will you find the following information about the machine?
   a. a diagram that shows its parts: _____
   b. how to take care of it: _____
   c. how to operate it: _____
   d. a troubleshooting chart: _____

6. Suppose you turn on the machine. The motor doesn't come on. What does the troubleshooting chart list as possible causes?
   _____
   _____
   _____

7. Suppose you check out each cause that's listed. None of them is the reason for the trouble. What would you do next?
   _____
   _____

Managing the Problem

Nadia opened her eyes again to look at the clock. 9:30 A.M. She had planned to get up earlier so that she could get a head start on the job ahead of her: tiling the bathroom floor. She and her husband, Tom, were expecting out-of-town guests. The bathroom floor, in the condition it was in, would be an embarrassment.

Nadia had suggested to Tom that they could save money by tiling the bathroom floor themselves. He challenged her to do it. After all, it was a small bathroom.

She saw herself going to the tile outlet and buying the tile. Surely three boxes of tile would take care of it, she thought. Maybe she should buy four boxes to be on the safe side. The store would probably buy back any unused tile. What else would she need? she wondered. Adhesive? Didn't they have a gallon in the basement left over from tiling the kitchen? But would it be enough? She decided that it would.

Nadia put on her jeans and a T-shirt, went into the bathroom, and surveyed it. Where would she begin? Around the sides or up the center of the floor? Maybe she'd have to start in an unnoticeable corner.

The tile store probably provided directions for laying the tile. She hoped so. She could hear Tom's advice ringing in her ears: "Hire a professional to make sure the job's done right." Anybody could lay tile, Nadia thought. She'd show Tom.

Later, at the tile outlet, Nadia picked out a pattern and color that she liked, only to find out that the pattern had been discontinued. She walked the display aisles for other ideas, but when she found a style and color that would go well in the bathroom she learned that it had to be ordered specially. Frustrated, she settled on the least expensive tile she could find. When she arrived at the checkout counter, she saw a sign that read: "Do you have everything you need? Have you forgotten your tile cutter, grout, chalk line, adhesive, and contouring tool?"

Nadia threw up her hands. All right. You win, Tom, she said to herself. She asked the store manager if he could recommend a professional installer. "Gladly," he said.

Nadia could hear Tom's voice saying, "I told you so."

Talk about It

- What does Nadia want to do?

- What is Nadia's problem?

- How could Nadia have prevented the problem from occurring?

Planning the Task

Sometimes a task itself may be a problem. The task may have several steps that need to be completed. For example, you read about the job Nadia set out to do: tile a bathroom floor. Her problem was that she failed to consider all of the steps needed to do a good job.

Handling a task that has many details can be stressful. Some people feel as if they are drowning when they think of everything that needs to get done. How do you usually feel when you must handle a task with several details?

One way to avoid trouble is to plan ahead. Come up with a solution for getting everything done *before* you start doing the task. The problem-solving strategy can help you make a successful plan of action.

▼ Work Out

Suppose you plan to paint a room. If you did each of these things before you did the task, would the task be easier to do? Tell why or why not.

1. Plan the steps for painting the room—figure how much paint you need, buy materials, prepare the room, and so on. Would that help ease your task? Why?

   _____

   _____

2. Decide what sequence in which to do the details. Would that help make your task easier? Why?

   _____

   _____

3. Make a list of all the materials—paints, brushes, pans, and so on—that you need. State which ones you have and which ones you need to buy. Would that help to make your task easier? Why?

   _____

   _____

Putting the Steps in Order

To handle a task with several steps, you need to organize yourself. Before you start the task, you should know:

- every step of the task
- what must be done to complete the steps
- the best order in which to handle the details

Here are some tips to help organize yourself.

Make a List

Begin by making a list of steps for the task. Write the steps down as you think of them. Don't think about how they should be done or in what order you should do them.

Order the Steps

You have done tasks with several steps before. Maybe you noticed that if you handle the steps in a certain order, managing and completing the task is easier.

You can sequence steps in several ways. One way is to consider how easy or hard it is to do each one. So you may decide to do all the easy steps or all the hard ones first.

Another way is to see if a task has a natural pattern to it. That is, certain steps must be done before others. For example, a certain step must be done first or last.

How do you know which way is best? You can't know right away. You need to read your list and think about the steps. You may find it helpful to imagine yourself doing each step. Soon you'll find yourself seeing how you should sequence them.

▶ When you have several chores to do at home, how do you organize yourself? (Easiest first? Room by room? etc.)

_____

_____

_____

Put the Steps in the Best Order

Once you know how you want to sequence the steps, you're ready. Carefully read your list and number the steps. Write *1* for the first step to do, *2* for the second, and so on. When you've finished, look over your sequence. See yourself performing the task. Does it make sense to you? If not, rearrange the order until it does.

When you actually do the task, you may find that your sequence doesn't work. Or you may see an even better order in which to do the steps. Be open to changing the order.

▶ Here is the list of steps for planting a garden. Read the list. Then put the steps in the order that you think makes the most sense. Number the steps. Remember, write *1* next to the first step to do, *2* next to the second, *3* next to the third, and so on.

Planting a Garden

_____ Find out what garden tools and materials such as fertilizer you need to buy or borrow.

_____ Prepare the soil for planting: plow it, add fertilizer if needed, and make rows.

_____ Learn what kinds of plants would grow best in the area and how to take care of them.

_____ Decide what vegetables to grow.

_____ Decide where you will plant each vegetable.

_____ Shop for seeds and seedlings, tools, and materials.

_____ Sow seeds or plant seedlings.

_____ Water plants.

You should have written the answers: 3, 6, 1, 2, 4, 5, 7, 8. Another correct order of steps might be: 4, 6, 1, 2, 3, 5, 7, 8.

In the case above, it's pretty clear that before you begin planting a garden, you first must learn what kinds of plants grow best in the area being considered. For example, a delicate vegetable such as lettuce, grown chiefly in warm climates like California, could not be expected to thrive in a colder state such as Minnesota. Once you have decided what plants grow best in an area, then you could decide where to plant them, you could shop for seeds, and so on.

▼ Work Out

Read the tasks in the two work situations below. Number the steps in the most sensible order to perform the task. One has been done for you.

1. A secretary's task is to get five copies of a report ready. She must complete the task in three days.

   The best sequence for the secretary to perform her task is:

   ____ Sort the copied pages into five report copies.

   ____ Type the report into her computer.

   _1_ Correct the report for grammar, spelling, and punctuation.

   ____ Punch holes in each copy and place each copy in a three-ring binder.

   ____ Proofread the report and make corrections.

   ____ Make five copies of every page in the report.

   ____ Create a title page and a table of contents.

   Why did you decide that this is the best sequence in which to do the task?

   _____

   _____

   _____

2. The last task for a florist is to close shop at the end of the day.

   The best sequence for the florist to perform his task is:

   ____ Count the day's income and record it in the books.

   ____ Count out change for the next day and put it in the store's safe.

   ____ Deposit money in the bank's outdoor automated teller machine.

   ____ Sweep the floor.

   ____ Bring in flowers from outside and store them.

   ____ Turn out the lights and lock the doors.

   ____ Make a list of flowers he needs to buy at the market tomorrow morning.

   Why did you decide that this is the best sequence in which to perform the task?

   _____

   _____

ON THE JOB

Suppose you're an office clerk. Your supervisor gives you this task: buy office supplies. You must buy the supplies tomorrow morning before you come into the office. The box below shows all the steps of your task. Read the steps; then answer the questions.

> **To Do**
>
> ___ Get supervisor's OK on typed purchase order.
> ___ Count the supplies that are left on the shelves.
> ___ Pay for order. Charge to office account.
> ___ Check store catalog for price of each item.
> ___ Drive to the office supply store in the morning.
> ___ Check inventory list to see how many pencils, pens, and other supplies the office should have.
> ___ Gather items in the store.
> ___ Type up purchase order.

1. What five steps must you complete today? Don't put them in any order.

   _____

   _____

   _____

2. What are two fact-finding questions that you might ask?

   _____

   _____

3. You'll need at least three sources to help you complete your task. What are they?

   _____

   _____

   _____

4. Study the steps. How would you most likely carry out the steps? Check your answer. Then explain why.

   _____ Do all the easy steps first.

   _____ Do all the hard steps first.

   _____ Follow the natural pattern of the steps.

   Why would you order the steps that way?

   _____

   _____

5. Now put the steps of the task in an order that you think would work best. Write a number next to each step in the box. Number them from one for the first step to eight for the last one.

Problem Solving in an Emergency

On Tuesday afternoon a forceful earthquake damaged the bridge. Ruben was stuck in the city for several days. He didn't mind. He was able to help out at the hospital where he worked.

But now Ruben was at home. He was trying to figure out how to carry out his daily routines. Today he had read bus and train schedules and leaflets and newspaper articles to help him find a way to and from work for the next six months. That's how long it would take to get the bridge fixed.

Before Tuesday, Ruben's commute to the city was easy. He got on a city bus at 7:00 A.M. He got off the bus right in front of the hospital at 8:15.

Ruben looked through his notes. He had several choices, but he didn't like any of them. He could take his usual bus. But since it now had to travel north to Vallejo Bridge to get across the bay, he would have to take the 5:30 A.M. bus.

He could take the 6:30 A.M. subway to the city. Then he would transfer two times on city buses to get to the hospital.

Or Ruben could drive down to the docks and take a ferry. A city bus from the ferry building went by the hospital.

During the San Francisco earthquake of 1989, commuters had to find alternate ways of getting to work.

But the fare for the ferry was three times the bus fare.

Ruben thought about driving his car. The hospital had a parking lot for its workers. But he couldn't stand the idea of driving in rush-hour traffic jams.

I will not get depressed, thought Ruben. He was lucky. This was the only problem that the earthquake had caused him. The earthquake had destroyed the home of his uncle's family. They were living in an emergency shelter.

Yes, Ruben thought. His transportation problem was nothing to complain about. Tomorrow he would try the subway.

Talk about It

- What problem is Ruben trying to solve?

- What did Ruben read to help him find a solution?

- Do you think solving problems in an emergency should be different from problem solving in normal times? Why or why not?

A Serious Situation

When an emergency happens, a solution must be found to handle the emergency right away.

Usually we think emergency situations result from accidents or disasters. For example, a car might collide with a train, or a hurricane might devastate a town. People, animals, and property may be hurt or in danger of becoming hurt.

In accidents and disasters people must think and act instantly to make themselves and others safe. They must solve problems very quickly. Some people, after the danger is over, are surprised that they came up with solutions.

But there are other kinds of emergency situations. Tasks or normal routines come to a stop because of a serious and sudden problem. Persons or things are not in any danger, but work cannot continue because of the interruption. In these situations a solution must be found as soon as possible.

Here are examples of crises that people face every day:

- A landlord has served an eviction notice to her tenants. They are allowed only thirty more days to live in their homes.

- Bus drivers have gone on strike. It may be a very long strike. People must now find other ways to get to work, school, stores, hospitals, and so on.

▼ Work Out

Think of an emergency that you had to handle. Answer the questions about it.

1. What was the situation?

_____

_____

2. What solution did you come up with?

_____

_____

3. Do you think the solution worked well? Why or why not?

_____

_____

Making the Best Choice

When you have two or more possible solutions, it often helps to think about the positives and negatives of the action you might take.

In the story on page 128, Ruben had to decide the best way to get to work. The list below shows the solutions he could choose from.

Ruben's Possible Solutions

- Catch the Innercity bus at 5:30 A.M. The bus would travel around the bay. The ride would take three hours.

  Round-Trip Fare: $3.00

- Catch a 6:45 A.M. ferry. In the city, catch a 7:05 A.M. bus. It takes about thirty minutes to get to the hospital.

  Round-Trip Fare: $8.50 for ferry, $2.60 for bus

- Catch a 6:30 A.M. subway. In the city, catch a 6:50 A.M. bus. Transfer at Polk and catch a 7:20 A.M. bus. Transfer at Geary and catch a 7:50 A.M. bus. It reaches the hospital at 8:25 A.M.

  Round-Trip Fare: $4.40 for subway, $2.60 for bus

- Drive own car. Leave house at 6:45 A.M. Drive around the bay. Get to hospital between 8:00 and 8:30 A.M.

  Total cost: $3.00 for bridge tolls, $3.00 for parking fees, and $3.00 for gasoline

---

Suppose Ruben's requirements are:

- Use public transportation.
- Pay only $10 for transportation each day.
- Begin traveling at 6:30 A.M.

Study Ruben's possible solutions and decide which one meets the three requirements.

You should have chosen "the *subway solution*." The subway solution meets all three of Ruben's requirements.

▼ Work Out

Below are people who live in Ruben's neighborhood and who work at the same hospital. Their solution choices are the same as Ruben's. (See the list, *Ruben's Possible Solutions*, on page 130.) Choose the best solution for each person's situation.

1. Mr. Fred Lloyd is an orderly at the hospital. His requirements are that he be at work by 8:30 A.M. and that he not travel on or below the water (BART, San Francisco's subway, travels through an underwater tunnel). What solution should he pick? Why?

   _____

   _____

   _____

2. Ms. Ching is a physical therapist. She works only three days at the hospital. Her requirements are that she not drive and that she spend two hours or less traveling. She wants to pay only $8 a day for transportation. What solution should she pick? Why?

   _____

   _____

   _____

3. Dr. Gomez works in the medical clinic. His requirements are that he get to his office between 7:30 and 8:00 A.M. and that he travel the shortest way (in travel time) possible. What solution should he pick? Why?

   _____

   _____

   _____

4. Ms. Julie Glover is a file clerk at the hospital. Her requirements are that she pay $5 or less for transportation and that she get to the hospital between 8:30 and 8:45 A.M. What solution should she pick? Why?

   _____

   _____

   _____

▼ Work Out

Minhee works in the shipping department of a publishing company. She gets to work at 8:30 A.M. She must complete these job orders. They are all urgent. Read the job orders and answer the questions that follow.

Job Order—Shipping Department

From: Customer Service Clerk
Date: March 5, 199___
    McKinley School must receive the attached order by March 16, 1990. Send third-class.

Note: Needs 25 copies of 5 different books. Must be ready for the 11:30 shipment. Take me 1 hr. to do.

Job Order—Shipping Department

From: President   Date: March 8, 199___
    The attached order must be received by March 10th. If necessary, send by overnight mail.

Note: Needs 100 copies of *Science*. Send express mail in the 11:30 shipment. Take me 40 min. to do.

Job Order—Shipping Department

From: Sales Manager
Date: March 9, 199___
    We need 2 copies of all special ed books for a conference today. Please have them ready by 11:30 for pickup.

Note: Needs 2 copies of 125 books. Take me 1 hr. to do.

---

1. What is Minhee's problem?

_____

_____

2. Check the facts that Minhee should consider finding to solve her problem. Write any other fact-finding questions that she should ask below.

_____ When was the request made?

_____ Who made the request?

_____ When must each order be completed?

_____ Who is to receive each order?

_____ How large is each order?

_____ How is each order to be shipped or picked up?

_____ How long will it take to fill each order?

_____ _____

3. Would these sources be helpful to Minhee? Explain why or why not.

    a. a map that shows where the books are in the warehouse

    _____

    _____

    b. the price list for the books

    _____

    _____

    c. procedures for overnight mail shipments

    _____

    _____

    d. special education book catalog

    _____

    _____

4. When Minhee gathers information, she writes a note to herself. Read her notes on each order. How do her notes help her complete the tasks?

_____

_____

_____

_____

ON THE JOB

Let's say you work at a youth center. Twelve young people who come to the center have a D average or are failing in school. The center decides to start a tutoring program.

All of you must work together on this problem:

What should the tutoring program be like?

1. Check the fact-finding questions that are most important. Write other questions that you can think of below.

_____ What subjects do the students have trouble in?

_____ What schools do the students attend?

_____ Where will the program get tutors?

_____ What hours should the tutoring program be open?

_____ How many hours long should a session be?

_____ _____

_____ _____

2. Check each source that would help you define the program.

_____ articles about tutoring programs that have worked

_____ the youths' school yearbooks

_____ the center's budget

_____ teachers and counselors in the youths' schools

3. Three possible solutions are suggested on page 135. What requirements would you consider when you decide on a solution?

_____ **a.** who should be in charge of the program

_____ **b.** cost of the program

_____ **c.** who should be tutors

_____ **d.** which tutors should work with which students

_____ **e.** when the tutoring sessions should be held

4. Suppose the program can have a budget of up to $3,000. Which do you think is the best solution: **A**, **B**, or **C**? Explain why you think that is the best one.

_____

_____

_____

_____

Possible Solutions

A. Get volunteers from high schools and colleges. Hold the sessions at the center from 3:00 to 8:00 P.M., Monday through Friday, and from 1:00 to 5:00 P.M. on Saturdays and Sundays.
Cost: $500

B. Use staff members and six youths from the center who have at least a B average as tutors. Youths would be paid $4 an hour. Hold the sessions at the center from 3:00 to 8:00, Monday through Thursday, and from 10:00 A.M. to 5:00 P.M. on Saturdays.
Cost: $2,800

C. Hire four professional tutors at $10 an hour. Sessions would be set between tutors and youth. The sessions should be held at the center but, if necessary, can be held elsewhere.
Cost: $2,000

You solve many kinds of problems in the workplace. Three common ones are figuring out why something stopped working, managing a task so that it's easier to do, and handling an emergency.

Using a strategy can help you examine a problem more clearly. In this unit you've learned how this problem-solving strategy can work for you.

Step 1: Define a problem.
Step 2: Make up a list of fact-finding questions.
Step 3: Gather your information.
Step 4: Analyze your information.
Step 5: Examine possible solutions.
Step 6: Choose the best solution.

One last word about problem solving. You don't have to wait for trouble to use your strategy on the job. Employers like problem solvers because they save the company money by finding ways to improve products and procedures. When you find a way to improve your job or workplace, think about telling your supervisor. Or, if your workplace has a suggestion box, think about writing your suggestions down.

UNIT CHECK

Explain what you do in each step of the problem-solving strategy.

1. Define the problem.

_____

_____

2. Make up a list of fact-finding questions.

_____

_____

3. Gather your information.

_____

_____

4. Analyze your information.

_____

_____

5. Examine possible solutions.

_____

_____

6. Choose the best solution.

_____

_____

TALK ABOUT IT

1. Do you think problem solving is best done alone or with other people?

2. When is the best time to report a problem to your work supervisor? Why?

3. Some people say that what happens in your personal life affects the way that you do your job. Do you agree or disagree? Why?

   a. Give examples of how problems in a person's personal life might cause problems on the job.

   b. Give examples of what a person can do to separate personal problems from the workplace.

FOR YOUR INFORMATION

Many inventions were the result of accidents. For example, the idea for stick-on notes came about when an employee accidentally spilled glue on a piece of notepaper. The idea for the microwave oven came about when a candy bar melted near a tube that discharged microwaves.

Comprehensive Review

This exercise gives you a chance to test your understanding of the key skills taught in this book. Complete the exercise and check your answers against the suggested answers in the key.

Let's say you work in a printing company. You get this memorandum from the company's president. Read the memorandum; then answer the questions.

Memorandum

Quality Printing and Copying House
To: All Employees
From: Mike Wingard
Date: Aug. 28, 199—
Re: SAFE! Program

    Within the last five years, each department at Q.P.C.H. has doubled. With the increase in <u>employees</u>, the rate of accidents has increased. We are lucky that so far none has been very serious. However, it is time to cut down such a high accident rate. Effective the first of September, Q.P.C.H. will start its new SAFE!—Safe Actions for Employees—Program. Debra Yuen will be its administrator.

    Throughout September, Ms. Yuen will lead safety workshops for each department. It is <u>mandatory</u> that all department <u>staff</u> attend their designated workshops. Ms. Yuen will be sending out workshop schedules next week.

    SAFE! has been designed in such a way that its success relies on complete employee participation. Accident prevention and investigations are a strong aspect of the program. With each accident, no matter how small, it must be determined whether it is unique—a one time happening—or one that could possibly occur to any worker in any department at any time.

    Effective today, Q.P.C.H. starts the following new procedure for reporting and investigating accidents.
1. As soon as possible after your accident, complete Form AR4— On-the-Job Accident Report. Be very specific in describing events that led up to the accident itself.
2. If others witnessed your accident, have them complete Form AR7—Accident Witness—and attach their forms to your report.
3. If you received medical attention, have the attendant complete Part B of Form AR4.
4. Give your report to your <u>supervisor</u>, who also will have completed his or her own report of your accident.
5. An investigative team made up of Ms. Yuen, your supervisor, you, and one employee from another department shall be formed to study the accident and recommend solutions to help prevent similar accidents from occurring.

    Please file this memorandum with your <u>employee</u> <u>handbook</u> until permanent section pages are distributed. Direct all questions or comments regarding the SAFE! program to Ms. Yuen.

Form M1-1980

A. Vocabulary p. 1–24

   1. The five terms below are underlined in the memorandum. In your own words, define each term. Use your context skills, if necessary.

   a. employees _____

   _____

   b. mandatory: _____

   _____

   c. staff: _____

   _____

   d. supervisor: _____

   _____

   e. employee handbook: _____

   _____

   2. Write *abbreviation* or *acronym* next to each job term below. Then write what the letters of each job term stand for.

   a. SAFE! is an _____.

   It means _____

   b. Q.P.C.H. is an _____.

   It means _____

   3. The company uses codes (letters and numbers) to identify its many forms. What is the code for the:

   a. memorandum form that the president used: _____

   b. On-the job new accident report form: _____

   c. accident witness report form: _____

B. Comprehension p. 25–48

   4. What is the main point of the memorandum?

   _____

   _____

   5. Give one detail that the SAFE! program involves.

   _____

   6. To whom would you address your questions about the SAFE! program?

   _____

   7. Where is each worker supposed to file the memorandum?

   _____

C. Following and Using Graphics p. 49–75

   8. Answer each question based on the flowchart.

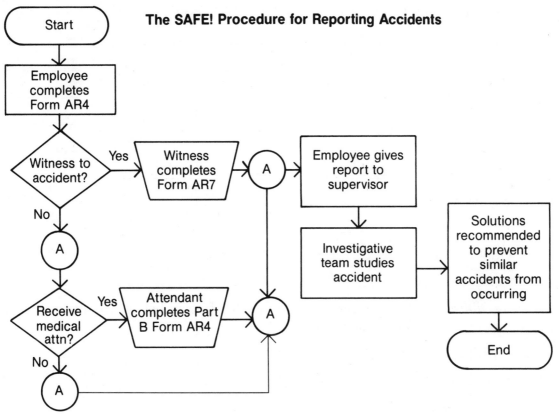

The SAFE! Procedure for Reporting Accidents

   a. If a victim of an unwitnessed accident who did *not* require medical treatment
      followed each step in the SAFE! procedure, how many steps would take place in
      the system? _____
   b. List each step.

   _____

   _____

   _____

   _____

   _____

   c. If a witness observes an accident, must he or she complete Part B of Form
      AR4? _____

   d. What is the final step in the SAFE! procedure?

   _____

   e. An employee had an accident, and a witness was present. No medical attention
      was given. How many steps took place to complete the procedure?
      _____

   f. One step that is included in the memo is *not* shown on the flowchart. What step
      is omitted?

   _____

D. Looking for Useful Information p. 93–105

9. Suppose you're skimming the table of contents of the employee handbook. You want information about accident reports. Which section would you most likely turn to?

_____ **a.** Safety in the Workplace

_____ **b.** Safe Work Habits

_____ **c.** When an Accident Does Happen

Why did you choose your selection?

_____

10. Suppose the following entry is in the employee handbook's index. You scan the entry to find information about the subjects below. Write the topic and page number that would most likely contain the information.

Accidents: Causes, 87. Injuries, 89, 100–101.
Investigations, 92–95. Medical reimbursements, 102.
Reporting, 90–91.
a. Filing an accident report: _____

b. Most common reasons for accidents: _____

E. Problem Solving p. 106–137

Let's say you are part of an investigative team. Read John Guzman's accident report; then answer the questions.

I was alone in the folding room. I was carrying a 50-pound box of brochures, which I was taking to the front office. I was walking by the worktables when I slipped on a brochure. I had not seen it at all. As I fell, I let go of my box, and it landed on my knee at the same time I landed on the floor. Tony Waters came in at that time and saw me fall. He called for help.

11. What caused the problem?

12. Give one fact-finding question that you would ask to help you analyze the problem.

13. Suppose you learn the following.

Workers store bad or extra pamphlets, brochures, etc., in piles beneath the worktables. It seems that one pile of brochures had slipped toward the aisle sometime before or at the time John Guzman slipped. John Guzman's accident is the only one thus far.

Would you say John's accident could or could not happen again? Why? _____

_____

14. Suppose your committee decides that this accident can occur again. What suggestion would you give to prevent such an accident from happening again?

_____

_____

_____

Answer Key

UNIT ONE: BUILDING A WORKING VOCABULARY

Lesson 1: Recognizing Terms

Page 2: Talk about It
- Cooks use precise words because each word describes a specific way of cutting and how the food should look.
- Other words that are commonly used in cooking include *sauté, puree, whip, steam, blanch.*
- It is important that you understand all the words that are used on a job so that you can perform your duties properly.

Page 3: Work Out: Coming to Terms
Across:
4. employer
6. supervisor
8. memorandum
9. co-worker
Down:
1. salary
2. department
3. personnel
5. policy
7. procedure

Page 4: Work Out: Special Job Terms
Answers will vary. Words should match the work environment in which they are used.

Page 5: Work Out: What the Terms Mean
2. c   6. d
3. a   7. h
4. g   8. e
5. b

Page 6: Work Out: Describing an Occupation
2. waiter or waitress
3. bank teller
4. shipping clerk
5. receptionist
6. salesclerk

Page 7: Work Out: Describing Tasks
2. Operate   4. Organize
3. Maintain   5. Assist
                        6. Perform

Page 8: Work Out: Knowing Your Job Terms
1. a, b   4. a, b, c
2. b   5. c
3. a   6. b

Page 9: On the Job
Answers will vary and will depend on students' experience.

Lesson 2: Reading "Shorthand"

Page 10: Talk about It
- KJ and Mel are looking for the National Aeronautics and Space Administration.
- They missed their sign because KJ didn't know that NASA stands for National Aeronautics and Space Administration.
- Answers may vary but should be similar to the following explanation. KJ has problems reading abbreviations and acronyms. It's important to be able to read abbreviations, because many jobs include them as part of a working vocabulary.

Page 11: Work Out: Reading Abbreviations
2. ASAP   4. misc.
3. F   5. ref.

Page 12: Work Out: Abbreviations for Phrases and Long Names
2. e   7. b
3. g   8. d
4. a   9. c
5. i   10. f
6. h

Page 13: Work Out: Reading Acronyms
2. a   8. l
3. f   9. h
4. e   10. i
5. b   11. g
6. c   12. j
7. k

Page 14: Work Out: Reading Symbols
2. h   6. g
3. a   7. d
4. f   8. c
5. b

Page 15: Work Out: Reading Symbols to Operate Machines

2. e   5. h
3. a   6. f
4. b

Page 17: On the Job
1. UPC
2. annual percentage rate
3. RFD
4. CETA
5. negotiable order of withdrawal
6. COLA
7. poison
8. %
9. copyright
10. to identify a product's type and price

Lesson 3: Understanding New Terms
Page 18: Talk about It
- Mr. Alvarez is attempting to fill out an application for a credit card.
- Lucy looks over the section and notices what kinds of information are being asked for.
- Yes, because Mr. Alvarez might have committed himself to paying for something he doesn't have to buy.
- Answers will vary.

Page 19: Work Out: The Context for a Word
1. Your answer should read something like "Make sure the names of all current files are recorded in alphabetical order on a list."
2. Answers will vary.

Page 21: Work Out: Finding a Definition
1. a   4. b
2. c   5. c
3. b

Page 22: Work Out: Choosing the Best Definition
Answers will vary according to the dictionary used. However, answers should be similar to the following: 1. the usual way of doing something 2. one-on-one 3. to ask for something 4. having to do with a patient's treatment

Page 23: On the Job
Answers will vary.

Page 24: Unit Check
Answers will vary for questions 1–5. For question 6 you can guess the meaning of an unfamiliar word by understanding how it is used in the sentence and by knowing the meanings of the words that surround it.

UNIT TWO: UNDERSTANDING WRITTEN IDEAS
Lesson 4: Reading to Fill Out Forms
Page 26: Talk about It
- Sheila gives her name, address, phone number, schools she attended, and skills she has.
- Companies ask the same questions to determine whether or not an applicant is the best person for the job they want to fill.
- Companies do not need to know an applicant's marital status and generally anything that has nothing to do with his or her ability to perform the job. Other answers may apply.

Page 27: Work Out: Filling Out Simple Forms
1. c
2. a, d

Page 30: Work Out: Subjects of Memos
2. 1, yes
3. 2, employee
4. 1, no
5. 1, no
6. 2, employee

Page 31: Work Out: Subjects of Memos
1. job evaluations
2. a. 2; b. 3; c. 1
3. the supervisor and the worker
4. worker's personnel folder and to the staff member
5. supervisor
6. The worker answers a list of questions.

Page 32: On the Job
1. the copy machine
2. **Paragraph 1:** who is in charge of the copy machine; **Paragraph 2:** how you can copy personal materials.
3. **Paragraph 1**
   a. Etta Lewis is in charge of the copy machine.
   b. Etta answers any questions or problems.
   c. Etta puts in more copy paper as needed.
   d. Maury Silva is Etta's backup.
   Paragraph 2
   a. Personal copies have to be paid for.
   b. Get supervisor's permission to copy personal papers.
   c. Personal copies cost 5¢ each.
   d. Pay Etta for personal copies.

Lesson 5: Getting the Main Point

Page 33: Talk about It
- Mrs. Glotov is complaining that her phone bill is too high for her to afford.
- Mr. Lucas is hoping that Mrs. Glotov will understand that she could apply for the phone company's program for senior citizens on a fixed income.
- Mr. Lucas could have made his idea clearer by getting to the point about the program when Mrs. Glotov first began complaining about the phone bill.

Page 34: Work Out: What's the Point?
2. a   **5.** b
3. f   **6.** d
4. e

Page 37: Work Out: Finding the Main Point
1. The subject is the format of a memorandum.
2. The main point of the first paragraph is what headings must be at the top of a memorandum.
3. The main point of the second paragraph is what to write after the headings in a memorandum.
4. Certain information must be the same on every memorandum.

Page 38: Work Out: Finding the Main Point
1. employee's handbook
2. **a.** Paragraph 1: what kinds of information are in the handbook
   b. Paragraph 2: changes will be made to the handbook now and then
   c. Paragraph 3: every worker is given a copy of the handbook
3. The employee's handbook gives workers information that they need to know about their job and the workplace.

Page 39: On the Job
1. ear drops
2. how to give ear drops to a child
3. **Paragraph 1:** how to prepare the ear drops
   Paragraph 2: how to give the drops
4. **Paragraph 1:** 1. Make sure you have the right drops.
   2. Read the prescription.
   3. Warm up the bottle of ear drops.
   Paragraph 2: 1. Bring the child to the first aid room.
   2. The child lies on his or her side.
   3. The drops should fall into the center of the ear.
   4. The dropper should not touch the child's ear.

Lesson 6: Reading Messages on Signs and Labels

Page 40: Talk about It
- Ruth and Bill disagree on the meaning of the sale sign.
- It seems that Ruth has the right understanding. Usually if a sale applies to all items, the store will post a sign that says that all such items are on sale.

Page 41: Work Out: What's the Message?
2. b   **4.** a
3. e   **5.** c

Page 43: Work Out: Interpreting Labels
2. To Be Done ASAP
3. To Be Filed
4. OUT
5. MAIL

Page 44: Work Out: Interpreting Signs
1. b
2. a

Page 45: Work Out: In the Workplace
1. equal opportunity
2. qualifications
3. physical handicap
4. discriminate

Page 46: Work Out: Understanding the Message
1. cigarette smoking
2. Smoke only in allowed places; don't smoke near computers or in the work area.
3. Smoking is not allowed in the work area.

Page 47: On the Job
Answers will vary.

Page 48: Unit Check
1. Three things that you can do to understand any memo are:
   a. Know the subject.
   b. Find the meaning of any terms you don't know.
   c. Make a note of the kinds of details that are given in each paragraph.
2. Identifying the main point of a text or sign helps you understand quickly what the writer is asking you to do.

UNIT THREE: MAKING GRAPHICS WORK FOR YOU

Lesson 7: A Picture Is Worth a Thousand Words

Page 50: Talk about It
- Laura studied diagrams that show how to change a tire.

- Laura wanted to look at the pictures instead of reading the words because the pictures were easier to understand when she was under the pressure of changing a flat tire.

Page 51: Work Out: Reading Pictures for Details

1. The jack handle is a long rod that is inserted into the jack.
2. The jack rises up when the handle is turned in the direction of the arrow.
3. You would turn the jack handle the other way or opposite the direction of the arrow.
4. In an emergency such as a flat tire a diagram is useful as a quick reference tool since a person generally doesn't have time to read instructions.

Page 52: Work Out: Seeing Connections

1. AC Adaptor
2. answering machine input
3. The phone would not work because the phone line is connected to the answering machine. When the machine has no power, the phone line is dead.
4. The phone would work. The power to the answering machine is still on although it's turned off.

Page 53: Work Out: Following Diagrams in the Workplace

1. a fire extinguisher
2. The main point is knowing your fire extinguisher or knowing the parts of your fire extinguisher.
3. pressure gauge
4. nozzle
5. b, c, e

Page 54: On the Job

1. The monitor, CPU, disk drive, keyboard, and printer
2. floppy disk
3. The monitor's function is to display the computer's functions and the information you enter.
4. The printer's function is to print out work that you've done on the computer.
5. CPU

Lesson 8: Finding Your Way

Page 55: Talk about It

- The nurse uses a map to show directions.
- Gilda can refer to the floor plan to help her find her way, so she doesn't have to keep the directions in her head.
- In cases of emergency, occupants of a building must be able to get out as fast as possible.

- Floor plans and maps can tell you the locations of places such as washrooms, bus terminals, and other places that can be represented by a symbol.

Page 57: Work Out: Getting Around with a Map

1. Go straight down the corridor that faces the elevator. The pharmacy door is on the left side.
2. Go straight down the corridor. At the end, turn left. The check-in counter is on the right side.
3. When you leave the gift shop, turn right. Turn at the second right. The optical lab is on the left side of that corridor.
4. When you leave the women's clinic, turn left. At the second corridor, turn right and walk to the end of it.

Page 59: Work Out: Finding Out What To Do

1. a. ▭   d. ◇
   b. ▢   e. ▽
   c. ◯
2. 5
3. No, the person repeats the driver's test.
4. 6
5. The clerk issues a temporary license.

Page 60: On the Job

1. delivering mail
2. Are all the departments identified on the mail?
3. 2
4. 4
5. sorting mail by department
6. department secretaries
7. Answers will vary.

Lesson 9: Getting Help from a Table

Page 61: Talk about It

- The salesclerk shows Josie a size chart or size table. It shows how big a baby must be for a certain size to fit.
- Josie would need the graphic to find out how the sizes of foreign-made clothes compare to U.S. sizes.

Page 63: Work Out: Finding Facts from a Table

1. At 9:00–10:30 A.M. on both days or at 7:00–10:00 on Thursday night
2. U.S. History, Civil Rights, or History of the Americas
3. Civil Rights or History of the Americas
4. No, because Conrad teaches only at night.

Page 65: Work Out: Solving Problems with a Table

1. The printer will not start.
2. The printer may have a burned-out fuse.
3. The problem might be one of these:
   a. The *Select* button isn't pressed.
   b. The printer cable is plugged into the wrong port.
   c. The ribbon has been installed incorrectly.

Page 66: Work Out: Solving Problems with a Table

1. immunization
2. The table shows the ages at which a child should get different immunizations.
3. diphtheria, whooping cough, tetanus, and polio
4. measles and German measles
5. measles, German measles, diphtheria, tetanus, TB, and polio
6. diphtheria, tetanus, and polio

Page 67: On the Job

1. To help me find out what action to take to correct the problem.
2. To water his soil thoroughly every now and then to wash the salt away.
3. The lawn may not root. Or the lawn may root, but it will die because the water won't drain.
4. The soil is too acidic.
5. Give the soil the right amount of fertilizer that has nitrogen in it.

Lesson 10: Seeing Changes over Time

Page 68: Talk about It

- Lily has tables, charts, and line graphs in her "Fat Binder."
- It motivated her to lose weight because she was able to see how much weight she was losing.
- If people can see how much progress they're making, it might give them greater motivation to stick to their goals.

Page 69: Work Out: Using a Line Graph

1. c
2. b
3. a
4. e
5. d

Page 71: Work Out: Reading a Line Graph

1. The subject of the graph is monthly sales of stuffed animals.
2. The main point is how much money the toy store took in from stuffed animal sales each month in 1990.
3. The store made the fewest sales in November.
4. The store made its greatest sales in December.
5. You might guess that people buy stuffed animals for Christmas gifts more than for any other occasion.

Page 73: Work Out: Comparing Two Trends

1. Except for the first month, computer A sales kept increasing during the time period.
2. Computer B sales were up and down during the time period.
3. Sales for both computers increased during March and June.
4. Computer A sales increased from February to June, while computer B sales went up and down.

Page 74: On the Job

1. The marks stand for each day that a person is ill.
2. Each long mark stands for one (1) degree.
3. a. hypothermia
   b. flu
4. A person's temperature increases rapidly the first four days. After it reaches the high point, it levels off and decreases rapidly to normal temperature.
5. A person's temperature decreases after the first day, and for several days goes up and down but still stays below normal temperature. By the ninth day, it reaches normal temperature (98.6°F) and holds for two days.

Page 75: Unit Check

1. table
2. diagram or schematic
3. map
4. line graph

UNIT FOUR: READING TO FOLLOW INSTRUCTIONS

Lesson 11: Reading and Following Instructions

Page 77: Talk about It

- He put too much soap in the washing machine, and suds overflowed.
- He could have read the instructions on the soap box.

Page 78: Work Out: Reading Instructions

1. a bottle of medicine
2. a credit card bill
3. a can of soup
4. a contest form

Page 80: Work Out: A Strategy for Following Instructions

1. The goal is to get a library card.

2. Fill out an application form, show proof of address, and pay a $2 fee.

3. Some proof of address such as a driver's license, a completed application form, and $2.

4. Answers will vary.

Page 82: Work Out: Applying the Strategy to the Job

1. The goal is to send out a letter in today's mail.

2. Type the letter, get a signature, make three copies, and mail the letter that day.

3. She needs letterhead, an envelope, and Mr. Burton's signature.

4. Answers will vary.

Page 83: Work Out: Applying the Strategy to the Job

1. Frank did not follow instructions correctly. He did not put side dishes, such as the salad and dessert, in separate bowls. He also did not put on an apron or plastic gloves.

2. Jenna did not follow instructions correctly. She placed their food on the tables before the participants were seated.

Page 84: On the Job

1. The goal is to get an order to a school by September 6th

2. You should have checked *Fill the order; Send the complete order by special delivery.*

3. You should have checked all three boxes.

4. Answers will vary.

5. You should have written something like "Ship the complete order for Willows School by September 4th by special delivery."

Lesson 12: Reading and Following Procedures

Page 85: Talk about It

● Bo is having a difficult time buying gasoline.

● He should have read the procedures for the gas station.

● Bo probably doesn't want to take the time to read new directions at every gas station.

Page 86: Work Out: The Goal

1. b   **2.** c

Page 87: Work Out: The Right Order

2 Take a transfer from the driver.

4 Pull the cord to signal that you'll get off at the next stop.

1 Put your fare in the collection box.

5 Get off the bus at the back door.

3 Find a seat for yourself.

Page 90: Work Out: Making a Flowchart

Title: How to Turn on a Computer

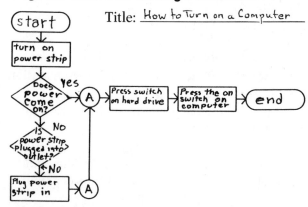

Page 91: On the Job
Answers will vary.

Page 92: Unit Check

1. The goal is to correct the tests and enter the results in the gradebook by tomorrow.

2. a. Correct math tests.
  b. Enter results in grade book.

3. tests, pen, grade book

4. Answers will vary.

UNIT FIVE: FINDING USEFUL INFORMATION

Lesson 13: Using a Reference Tool

Page 94: Talk about It

● Roy was looking for the page advertising camping gear.

● Roy used the index to find what he was looking for.

● You use the index, table of contents, or tabs on pages to find specific information in a catalog, manual, or other book.

Page 95: Work Out: Using The Table of Contents

2. Education Information

3. Libraries

4. Restaurants

5. Shopping Areas or Public Transportation

Page 97: Work Out: Where Do You Find It?

1. Word processing systems

2. Wages or Salary

3. Word processing function keys

4. Word processing equipment, care of

Page 98: Work Out: Parts of a Manual

Across:
3. Title
5. Contents
6. Copyright
7. Index
8. Appendix

Down:
1. Illustrations
2. Glossary
4. Introduction

Page 99: On the Job
1. Who's Who in the Company
2. Workers' Status and Pay
3. Safety Rules
4. You'd probably look under *Work Rules.*

Lesson 14: Reading Quickly for Information

Page 100: Talk about It
- Marva took her time in reading personal letters.
- She probably read her personal mail more carefully than the rest because she could get a general idea of what the other mail was about by scanning it.
- Marva organized the mail into piles and read each kind of pile in a different way.
- Answers will vary.

Page 101: Work Out: Scanning or Skimming?
1. You would skim the article, because you want a general idea of what is being said.
2. You would scan the recipe, because you want a specific fact.
3. **a.** You would scan the ads, because you are looking specifically for ads aimed at part-time office workers.
   b. You would skim the ad, because you are getting a general idea of what is being said.

Page 103: Work Out: Skimming through Text
1. b
2. Answers will vary but should include the idea that an enumerator may make address lists, distribute and collect census forms, or interview people.
3. Enumerators go out to the places that people live.
4. Answers will vary.

Page 104: On the Job
1. The main point of this memo is: Employees must follow a new procedure when persons visit them at the company.
2. The new procedures start on October 1, 199___.
3. The receptionist must call someone such as the secretary in the accounting department and tell him or her that the service person is there. She must write down the time the service person entered and left the building.
4. You take the visitor back to the lobby.
5. You could not take her on a tour at the moment, because you need to give at least two days' notice that you're taking someone on a tour.

Page 105: Unit Check
1. The table of contents can help you get an overview of what's in a book.
2. The index can help you find the pages where you would need to scan for the information that you need.

UNIT SIX: PROBLEM SOLVING IN LIFE
Lesson 15: A Problem-Solving Strategy

Page 107: Talk about It
- Tania is having trouble managing her money.
- Imagine that the family is a business and come up with a plan to get out of debt.
- Making a plan can help people organize ideas so that they can see what makes sense and what may work.

Page 110–111: Work Out: Finding a Solution
1. All the choices should be checked. You could also add these questions: What could be done by a part-time worker? What tasks are the easiest and hardest to train someone to do? What does Fran most enjoy doing herself?
2. a secretary's handbook; office procedures; past job orders; and customer invoices; office phone log; sales reports
3. Answers will vary but may be similar to these:
   a. The part-time worker should do mostly the least difficult tasks since he or she would be in the office for only a few hours each day. The worker could answer phones and greet visitors, file, copy, and run errands.
   b. The full-time worker can help Fran do the more difficult tasks as well as fill in when the part-time worker is not available to take over the receptionist's tasks.

Page 112: On the Job
1. How many books does the library have now? How many books does the library expect to get each year? How many books fit on a shelf? What is the cost of a shelf? When would the library have funds to buy shelves? Are there any books that could be stored right now?
2. book inventory reports, circulation reports indicating how often books are checked out, lists that state what books are on order, current budget report, price lists for shelves

3. a. Some possible solutions are: (1) Remove all the books that are rarely borrowed and store them. The library won't need to buy shelves for a few years. (2) Remove at least 2,000 old books and store them. The library can buy new shelves as soon as possible. (3) Find the money to buy at least five shelves for only new books. Place the shelves in the main part of the library.

b. Answers will vary.

Lesson 16: Finding the Cause of a Problem

Page 113: Talk about It
- The vacuum cleaner isn't picking up dust.
- The machine's air outlet was blocked.
- The strategy would have helped her focus so that she could find the cause of her problem sooner.

Page 114: Work Out: When Something Stops Working
Answers will vary.

Page 116: Work Out: Understanding How Something Works
1. As air moves into the machine, it carries dust and other things along with it.
2. Air enters through the machine's nozzle, travels up through the hose, rises through the dust bag and air filter into the area where the fan is, and then leaves through the air outlet.
3. Air could not move through the machine, so it would not work.
4. No. Less air would be able to travel through the machine.

Page 119: Work Out: Locating the Problem
1. Answers will vary. Possible answers include: The fuse controlling the outlet could be out; Mr. Torres didn't turn the power on correctly.
2. nozzle, hose, fan, and air outlet.
3.

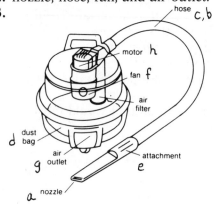

4. **a.** Repair or replace it.
   b. Clean it out.
   c. Replace it.
   d. Repair or replace it.

Page 121: On the Job
Answers will vary.

Lesson 17: Managing the Problem

Page 122: Talk about It
- Nadia plans to tile the bathroom floor.
- Nadia's problem is that she failed to plan ahead for the task.
- Nadia could have planned how she would do the task before she started doing it.

Page 123: Work Out: Planning the Task
1. By figuring out how much paint you need, buying the materials, and preparing the room, you will not have to waste time once the job is started. Also, by preparing the room properly you won't need to redo the job.
2. By deciding the order in which you want to do the job, you make the task easier by knowing ahead of time what you will need to do when.
3. By listing all the materials you will need, you will make your job easier because you won't spend money on things you don't need and because you'll buy things that are necessary for the task ahead of time.

Page 126: Work Out: Put the Steps in the Best Order
1. You should have decided on this order for the steps: 6, 2, 1, 7, 4, 5, 3. The main point is that the secretary's job has only one logical sequence. The steps must follow a sensible order.
2. Answers will vary. The main point is that the florist's task has several possible sequences. No one order is necessarily better than another. However, depositing the money in the bank would probably be the last step.

Page 127: On the Job

1. Get supervisor's OK on typed purchase order; count the supplies that are left on the shelves; check store catalog for price of each item; check inventory list to see how many pencils, pens, and other supplies the office should have; and type up purchase order.

2. Some questions that you might ask: Where will I buy the supplies? How much do I have to spend? Where is the store? When does it open? How do I make a purchase order?

3. office supply inventory list, store catalog, and purchase order or purchase order procedures

4. The steps have a natural pattern to them. Certain things can be done only in the office or in the store. Some steps must be done before others, such as counting the supplies on the shelves before typing up a purchase order.

5. You should have decided on this order for the steps: 5, 2, 8, 3, 6, 1, 7, 4. The only variation in this pattern would be between 1 and 2. You could count supplies first and then check inventory instead of vice versa.

Lesson 18: Problem Solving in an Emergency

Page 128: Talk about It

- Ruben is trying to decide what is the best way to commute to and from work for the next six months.

- He read materials that gave him information about different forms of transportation and their time schedules.

- Problem solving during an emergency may need to be done more quickly than in normal times, but the steps should be followed in the same way.

Page 129: Work Out: A Serious Situation
Answers will vary.

Page 131: Work Out: Making the Best Choice
1. Fred's best solution is to drive his car or take the bus. That is the only way he would be able to get to work by 8:30 A.M.

2. Ms. Ching's best solution is to take the subway. That commute would take about two hours and would cost her $7 a day.

3. Dr. Gomez's best solution is to take the ferry. That's the shortest route of all four choices and the only one that would get him to the hospital before 8:00 A.M.

4. Julie's best solution is to take the Innercity bus. That gets her to the hospital by 8:30 and costs less than $5 a day.

Page 132–133: Work Out: Making the Best Choice

1. Minhee must fill three urgent orders by 11:30 A.M.

2. Answers will vary; however you should be able to support your choices with sound reasoning.

3. Answers will vary; however you should be able to support your choices with sound reasoning.

4. Answers will vary; however, you should have written that Minhee's notes will help her plan the tasks by estimating how much time it will take her to do them. This planning will help her use her time effectively.

Page 134–135: On the Job

1. Answers will vary, but you should be able to support your choices with sound reasoning.

2. The following should be checked: articles about tutoring programs that have worked; the center's budget; and teachers and counselors in the youths' schools.

3. a, b, c, e

4. Answers will vary.

Page 136–137: Unit Check

1. Describe the problem as clearly as possible.

2. Think up questions to ask to get facts that may help you solve the problem.

3. Find the facts in written materials, from people, and from other sources that you think may have the information.

4. Explain what all the facts mean.

5. Think up possible solutions that may solve the problem.

6. Choose the one solution that would most likely work out.

COMPREHENSIVE REVIEW

Pages 138–141

Page 139:

A. **Vocabulary**
   Wording for answers will vary.
   1. **a.** persons who are paid to work at Q.P.C.H.
      b. required; necessary
      c. all persons who work in a specific department at Q.P.C.H.
      d. worker who oversees you on your job
      e. a guidebook that describes all the rules, programs, and other important information that an employee at Q.P.C.H. would need to know
   2. **a.** SAFE! is an acronym. It means Safe Actions for Employees (Program).
      b. Q.P.C.H. is an abbreviation. It means Quality Printing and Copying House.
   3. **a.** M1-1980   **b.** AR4   **c.** AR7

Page 139:

B. Comprehension
   4. The company is starting a new safety program to cut down the high accident rate.
   5. Either of the following answers is correct: safety workshops for each department; a new accident reporting and investigating procedure.
   6. Debra Yuen
   7. in his or her employee handbook

Page 140:

C. Following and Using Graphics
   8. **a.** A victim of an unwitnessed accident who didn't require medical treatment would follow four steps in the procedure.
      b. The steps are:
         1. A form AR4 is completed.
         2. The form is given to the supervisor.
         3. An investigative team studies the accident.

   4. Solutions are recommended to prevent similar accidents from occurring.
   c. No. A witness who observes the accident completes Form AR7.
   d. The final step in the SAFE! procedure is recommending solutions to prevent similar accidents from occurring.
   e. The victim of an accident in which a witness was present and no medical attention was given must go through six steps to complete the SAFE! procedure.
   f. The step that is omitted is: The supervisor completes his or her own report of the accident.

Page 140:

D. Useful Information.
   9. **c.** When an Accident Does Happen. You should have chosen this selection because the title describes what to do in the case of an accident.
   10. **a.** Reporting, pages 90-91 **b.** Causes, page 87

Page 141:

E. Problem Solving
   11. There was a loose brochure on the floor.
   12. Some questions you might ask are: Where did the brochure come from? Are there other loose brochures in the area? Could John have prevented the accident in any way? Why wasn't John able to see the brochure on the floor?
   13. The accident could occur again because there's nothing to prevent the piles from slipping and getting in the way of workers as they are walking back and forth.
   14. Possible suggestions: Store brochures in boxes beneath the tables. Throw out all extra papers immediately so that there is no clutter. Store brochures in boxes away from the aisles.